OTHER TITLES OF INTEREST FROM ST. LUCIE PRESS

The ISO 14000 14000 EMS Audit Handbook

Gregory P. Johnson

S_L^t

St. Lucie Press
Boca Raton, Florida

Copyright ©1997 by CRC Press LLC
St. Lucie Press is an imprint of CRC Press

Printed and bound in the U.S.A. Printed on acid-free paper.
10 9 8 7 6 5 4 3 2 1

ISBN 1-57444-069-1

Phone: (561) 994-0555
E-mail: information@slpress.com
Web site: http://www.slpress.com

S^t_L

Published by
St. Lucie Press
2000 Corporate Blvd., N.W.
Boca Raton, FL 33431-9868

TO DAD
The man I admire and respect more
than anyone else in the world....
Thanks for "the book the house
built, Part II."

Table of Contents

Preface

How one gets involved initially in environmental auditing is usually a pretty interesting story. I was in a position in the early 1980s to continue down the "quality" road, and, quite frankly, never gave much thought to the environment in general. Then one day (we'll say somewhere in America) during a pre-award inspection system audit, I made a conscious decision to ask questions outside the scope of my audit, as a rather creatively decorated 55-gallon drum caught my eye (painted on the drum was the "Grateful Dead" skull and crossbones logo). I inquired, "What's the material in this drum here?"

The supervisor responded, "Oh, that's just some hazardous by-product material from our manufacturing operations."

"What do you do with it from here?" I asked, having no idea that the answer would eventually change my professional life.

The supervisor cautiously looked to his left, then his right, and said, "Well, we have this carrier who, with no questions asked, takes this stuff off our hands every other week. We pay him in cash and from there, I have no idea where it goes."

Hmm...this can't be right.

To make an incredibly long story short, several companies were shut down, several people indicted, and I was thrust into the world of environmental auditing. Of course, by mistake.

From there, I was "encouraged" to participate in a variety of environmental site assessments for several aircraft, aerospace, and defense contractors on the West Coast. For several years, I juggled the home life with extensive environmental audits of all types from coast to coast, while taking evening and weekend correspondence and college classes on every "environmental issue and regulatory legislation imaginable." For several years, I worked directly with and for what I could only describe as "pathetic and border-line" illegal audit programs. (Hey, they were signing my paycheck, so what did I care?)

Well, I did care…and sometimes worried. Enough so that after extensive experience and training in the world of environmental auditing, I designed and developed a five-day training course titled "Environmental Quality Audit Techniques for Commercial, Government and Nuclear Industries" in early 1988. Seven years later after over 70 classes and 800+ extremely knowledgeable participants, I started hearing rumors about ISO 14000. Another conscious decision was made to direct all of our training organization's efforts into this area.

The rest, as they say, is history and brings us to present day.

Acknowledgments

The reader should be advised that all text marked with ◆ is reproduced from ISO standards under International Organization for Standardization, ISO, Copyright License number Q.S.D./1CC/1997. No part of these ISO standards may be reproduced in any form, electronic retrieval system or otherwise, without the prior written consent of the International Organization for Standardization, ISO, Case postale 56, 1211 Geneva 20, Switzerland, Fax: +41 22 734 10 79, E-mail: central@ISO.CH or of the American National Standards Institute, ANSI, 11 West 42nd Street, 13th floor, New York, N.Y. 10036, USA, Tel: +1 212 642 4900, Fax: +1 212 398 0023.

The list of people to thank and contributors to this book is endless. First of all, this book could not have been developed with the contributions of each and every participant that ever attended our compliance and/or EMS auditors courses around the country. The interaction, workshops, and extensive comments provided in these courses was, in itself, the greatest learning process of my career.

Hats off to all of you….

To Mr. Anton Camarota, for his assistance in the initial development of the Environmental Quality Auditors Course and for pointing me in the right direction regarding the additional training I needed at my environmental infancy…. Thanks, buddy.

To Mr. Earl Roper, for giving me my first shot at quality systems auditing. A man I greatly respect and admire. It's hard to say where I'd be today without his initial confidence in my abilities…. Thank you.

Our ISO 9000 staff and instructors at Quality Systems Development (Q.S.D.), who, above all, kept the business going and kept me in line when I got too excited about things. Truly experts in crisis management and its effective implementation….

It has been a real honor to be associated with these extremely knowledgeable and patient individuals, including John Thornton, Steve Moubray, and

Pat Hanes. My best wishes for you all and nothing short of continued success and happiness in everything you do....

To Nancy Hall, technical editor, advisor, and proofreader of this book in its entirety, including various marketing and course material for Q.S.D. What an honor it has been to work with an environmental scientist and auditor throughout the development of this book and beyond. (Note to readers: You probably would not be reading this now if it wasn't for Nancy's endless efforts and "burning of the midnight candle." A world of thanks to you, Nancy!)

To the current and former ISO 14000 staff and instructors, including Michael Ross, EMS President of the Registrar Accreditation Board. Mike is one of the most active members of the U.S. TAG/TC 207 in the development of the ISO 14000 series standards. His input, advice, and representation of Q.S.D. has been a catalyst and the true backbone of providing up-to-date information on ISO 14000 issues around the world. I certainly look forward to our long, successful relationship with ISO 14000 and beyond. Kudos to you, my friend.

To Donald A. Sayre, our theatrical representative extraordinaire.... His polished skills in course development and presentation have gone well beyond the call of duty. Don's endless humor and enthusiasm made all of this worthwhile, including his extensive contributions and proofreading of this book. A true colleague, associate, and friend.

To Greg Hale, former editor of CEEM's *International Environmental System Update* and now with the Global Environmental Technology Foundation (GETF). I have nicknamed him the "Elliot Ness" of ISO 14000.... Thank you for the hours and hours of conversation, advice, and direction on all related ISO 14000 issues. I look forward to our long-lasting relationship (and watch out for the "crank" E-mail you were promised!)

A special thanks must go out to Brookes Cook and all the hard-working folks at CEEM, publisher of the ISO 14000 *International Environmental Systems Update* (IESU). This publication is dedicated to addressing anything and everything on ISO 14000, and anyone involved in the development, implementation, and maintenance of ISO 14001 would have an incomplete set of resources and information without it...including myself and our staff!

Many contributions from the IESU are included in this book. A special thanks to the individuals whose original works made their way into this book, including Ron Black, Thomas P. Ambrose, Marilyn Block, Robert Ferrone, Jean McCreary, Roger Brockway, and Cornelius (Bud) Smith, all of whom are highly seasoned professionals and experts in ISO 14000-related issues. Remember the names...they may be great sources of additional information, service, and training to you, the reader, in the future.

To my wife, Kelly, for standing by me through thick and thin, for her tireless effort and involvement in Q.S.D., and for being the kind of "partner" that one can only dream about. Thanks, hon (note: She's kissing me as I'm writing this!)

To my daughter, Hailey, for understanding why Dad's gone all the time (sort of), for keeping the volume down on her Alannis Morrisette tape while I'm working, and for all her help with the inspection of our course materials and advertising at Q.S.D. Destined to be the third generation of Johnson involved in management systems auditing.... Thanks, sweetie.

To Mom...for believing in me in everything I do.... Her support and advice truly led me in the right direction, personally and professionally.

And finally, there is L. Marvin Johnson...DAD....world pioneer in quality management systems auditor training and author of *Quality Assurance Programme Evaluation*, the first book ever written in the world on extensive management systems auditing.

I can look back now with a great deal of appreciation for growing up in his household. I understand why he had me proofread course materials, advertisements, and various revisions of his book time and time again, starting from the age of nine. I'm not sure if it was a master plan or what, but everything I did for his company had some relevance to auditing.

Many of his concepts, techniques, and applications of the audit process have been incorporated, modified for environmental applications, and included throughout this book. His endless support, advice, direction, and true appreciation for "doing it right the first time" have molded my abilities, knowledge, and dedication to auditing into what they are today. Thank you, my father, for your patience, guidance, and understanding in all we've been involved with.

A final thanks to you, the reader, for your interest and patronage. My true wish is that you may take from this book some concepts and techniques in EMS auditing and continuously improve your organization's efforts in making the audit process the best it can be...and then some. I encourage your input and comments regarding the information concepts and techniques presented in this book. Please feel free to contact me directly at Quality Systems Development, 105 Woodside Court, Leitchfield, Kentucky 42754 (tele: 502-257-9997, fax: 502-257-2740).

1

Introduction to EMS Audits

What Is an Environmental Management System?[1]

An environmental management system (EMS) is that aspect of an organization's overall management structure which addresses the immediate and long-term impact of its products, services, and processes on the environment. It provides order and consistency in organizational methodologies through the allocation of resources, assignment of responsibilities, and ongoing evaluation of practices, procedures, and processes.

An EMS is essential to an organization's ability to anticipate and meet growing environmental performance expectations and to ensure ongoing compliance with national and international requirements. EMSs succeed best when corporations make environmental management among their highest priorities.

In general, environmental management systems should achieve the following principles:

- establish an appropriate environmental policy, including a commitment to prevention of pollution;
- determine the legislative requirements and environmental aspects associated with the organization's activities, products, and services;
- develop management and employee commitment to the protection of the environment with clear assignment of accountability and responsibility;
- encourage environmental planning throughout the full range of the organization's activities, from raw material acquisition through product distribution;
- establish a disciplined management process for achieving targeted performance levels;

1

- provide appropriate and sufficient resources, including training, to achieve targeted performance levels on an ongoing basis;
- establish and maintain an emergency preparedness and response program;
- establish a system of operational control and maintenance of the program to ensure continuing high levels of system performance;
- evaluate environmental performance against the policy and appropriate objectives and targets and seek improvement where appropriate;
- establish a management process to review and audit the EMS and to identify opportunities for improvement of the system and resulting environmental performance;
- establish and maintain appropriate communications with internal and external interested parties; and
- encourage contractors and suppliers to establish an EMS.

The EMS Movement[2]

Public concern over industry's impact on the world's environment is increasing. Politically oriented bodies such as environmental advocacy organizations, watchdog groups, and the "green" parties that have established footholds in most European parliaments are urging businesses to take responsibility for their environmental effects. This pressure from the public sector has led to a rash of proposed and enacted environmental legislation worldwide.

However, recent reports are showing that companies choose to implement an EMS far more for internal management system efficiencies, waste reduction, and proactive regulatory compliance than for any other purpose.

One survey that supports these reports is the *Voluntary Environmental Audit Survey of U.S. Businesses* published in April 1995 by Price Waterhouse LLP. It found that of 369 respondents, 75% have an environmental compliance auditing program in place. While the survey did not address EMS audits specifically, it found that fully 96% of those that audit do so because "problems can be identified internally and corrected before they are discovered by an agency inspection." Another 94% stated that they audit "to improve our company's overall environmental program and make it proactive."

The reasons many companies are now adopting an environmental management system include the following:

- improve compliance with legislative and regulatory requirements, including requirements that certain information relating to environmental performance be made public;

- reduce liability/risk;

- generate a desire to benefit from regulatory incentives that reward companies showing environmental leadership through certified compliance with an EMS;

- prevent pollution and reduce waste;

- generate a desire to profit in the market for "green" products;

- improve internal management methods;

- manage pressure from shareholder groups;

- create community goodwill;

- generate interest in attracting a high-quality work force; and

- manage insurance companies' unwillingness to issue coverage for pollution incidents unless the firm requesting coverage has a proven environmental management system in place.

Development of the International EMS Standard[3]

Efforts to create a single, generic, internationally recognized EMS standard have been driven by the desire among companies to avoid duplicate— and sometimes competitive—corporate and governmental programs and by their need for objective validation of their commitment.

Such efforts have begun on several fronts, but one has led the way to ISO 14001. In England, the British Standards Institution (BSI) developed BS 7750, Environmental Management Systems, as a companion to its BS 5750 standard on quality management systems. BS 5750 was the forerunner and template for ISO 9000. In 1991, the Strategic Advisory Group on Environment (SAGE) was established by the International Organization for Standardization (ISO) to make recommendations regarding international standards for the environment, and it spent nearly two years studying BS 7750 and other national EMS standards to determine the need for an ISO international standard. The result of its study was the formation of Technical Committee (TC) 207 and the beginning of the development of ISO 14001.

Introduction to EMS Audits

Involved management and discipline is the key to environmental quality. Audits are the investigations that determine the extent of an activity's ability to implement and maintain an effective EMS. EMS audits are performed by contractors, customers, environmental and quality personnel, and government and regulatory agencies. Teams are composed of combinations of technical and administrative personnel, seldom with sufficient continuing experience to perform this function in the highly professional manner required. Therefore, an experienced and qualified auditor must lead a group of technical or admin-istrative individuals in any audit situation. When formal training cannot be provided for these individuals, it becomes the responsibility of the experi-enced auditor to ensure there is adequate performance of the audit function. Without this guidance, the data collected, the conclusions, and the recommen-dations may be suspect and even invalid.

Audits to monitor and measure environmental systems performance pro-mote close coordination and develop mutual understanding. Most regulatory and government agencies responsible for complex systems and projects audit contractors and their major subcontractors. In addition, hundreds of contrac-tors are administered by the Environmental Protection Agency (EPA) and the Department of Energy (DOE). They continually audit and review the environ-mental performance of contractors within their regulatory jurisdiction.

Training and preparations for performing audits in some industries have been self-induced. Until about 1980, those who had performed environmental program audits taught themselves from the knowledge they had gained through working in quality control and related environmental fields. Little effort was expended by major manufacturers, or even the government, for that matter, to train and qualify environmental audit personnel.

Since the author began intensified environmental auditor training courses for commercial and government industries in 1988, there has been increasing emphasis on "qualifying" individuals through seminars. Standards used to qualify these individuals included the following:

- ANSI-NQA-1 2S-3, "Supplementary Requirement for the Qualification of Quality Assurance Program Personnel"
- ANSI-N45.2.23, "Qualifications of Quality Assurance Program Audit Personnel for Nuclear Power Plants"

Quality Assurance Program Personnel? Isn't this a book on Environ-mental Management Systems Audit?

Yes, indeed, it is.

However, until the Environmental Auditors Registration Association (EARA) came along in 1994, there were no recognized qualification/certifi-cation criteria set forth for environmental auditors.

EARA is a UK-based global organization that registers environmental auditors to conduct EMS audits for ISO 14001, BS 7750, and the European Union's Eco-Management and Audit Scheme (EMAS). EARA also provides accreditation to Foundation and Advanced EMS audit course providers. The author's organization was the first company in the world to become an accredited Advanced EMS audit course provider for quality and environmental professionals specifically tailored to the ISO 14000 series draft audit standards.

The author's conclusion, based on personal observation, was that 40–50% of the people performing environmental program or compliance audits were not adequately trained nor qualified, nor did they have the knowledge to conduct investigations. Great strides have been accomplished in the auditor qualification/registration and training functions since 1988. However, those who are performing any type of environmental audit/assessment without adequate training are probably doing more harm than good. One might inquire, "What is adequate?" The following text presents, in considerable detail, material that the author believes is essential to performance of successful EMS or compliance audits.

The auditor requires extensive knowledge of quality management, environmental principles, a number of discreet probing techniques, and a conscientious desire to obtain facts. In addition, gaining proficiency also requires *time, training, and experience.*

Which brings us to you, the reader. To acquire the true spirit of the intent, concepts, and methodologies of EMS auditing presented in this book, let us explore several types of qualified auditors that are with us at this time.

Quality Systems Auditors (QSA)

It is the author's feeling that many organizations' management will consider qualifying quality systems auditors to conduct EMS audits. And rightfully so! However, many of the concepts and methodologies for successful audit of the EMS requirements set forth by ISO 14001 may appear very foreign to the QSA. While some of the requirements of ISO 14001 are similar to ISO 9001 requirements (see Chapter 5), the following are distinctly different from quality systems audit requirements:

- environmental aspects;
- legal and other requirements;
- objectives and targets;
- environmental management programme(s);
- communications;
- emergency preparedness and response; and
- monitoring and measurement.

Without a strong environmental background in the sciences, technology, environmental legislation, analytical laboratory operations, and environmental performance issues and principles specific to the facilities' operations, it is questionable how effective the QSA will be in auditing these requirements. The EMS audit organization should consider QSAs for the following requirements:

- structure and responsibility;
- training, awareness, and competence;
- environmental management system documentation;
- document control;
- nonconformance and corrective and preventive action; and
- records, environmental management systems audit, and management review.

Environmental Compliance Auditors (ECA)

One of the most critical messages to get across to the ECA, based on personal observation and discussion with the ECAs who have attended the author's EMS audit courses, is:

**AN EMS AUDIT IS NOT
A COMPLIANCE AUDIT!**
Never was.
Never will be.
<u>And</u> *there is no effective process for auditing
total compliance to both*
AT THE SAME TIME!

It is quite common in the author's EMS training courses that the ECA will take off their EMS auditor's hat and immediately put on their ECA hat when the potential of a regulatory noncompliance is detected in our case study workshops. This changes the investigative approaches and questioning techniques pre-developed by the EMS auditor. In addition, the true scope, purpose, intent, *responsibility, and authority* of the EMS audit may no longer be carried out as designed. Chapters 5 and 12 cover in detail what an EMS auditor should do if a potential regulatory noncompliance surfaces in the EMS audit.

The adequately trained ECA, after obtaining additional qualification in EMS audit principles, maybe the most effective person to consider for the EMS audit process. A thorough knowledge and background in the *distinctly different* requirements and technical qualifications expressed on the previous page (see QSA) are preferred to ensure the effective execution of the EMS audit processes.

Management

The author is very happy to have you with us…if the author had one wish specific to an organization's EMS and their audit programme, it would be (with no sales pitch intended) for all responsible managers, with defined authority within EMS, to read this book and/or attend an Advanced EMS audit course.

Unfortunately, in the author's opinion, there is probably a better chance at winning the lottery…twice. *The most common complaint of over 2,500 trained quality and environmental management system auditors over the past ten years is:*

> *We rarely have enough time to do everything we need to in conducting the audit.*

Additionally, the #1 reason for this appears to be:

> *Management does not allocate sufficient resources, personnel, or time to effectively conduct the audit.*

If you are saying to yourself at this point, "That is obviously not true!", we commend you.

You are obviously on the right track.

But it is interesting to note the direction many of the "gurus" of systems auditing have taken lately. See if you recognize any of the following current "buzz" terms relevant to systems auditing:

- value-added
- performance-based audits
- audit for continuous improvement

Value-added? This "buzz" term really cracks me up. Are these "gurus" insinuating that management is not aware of the "value" of management systems or performance or compliance-based audits?

Ladies and gentleman, these concepts are basically no different from the original and founding works put together by L. Marvin Johnson in his book entitled *Quality Assurance Program Evaluation*, first printed and published in 1970.

But again, what might these gurus be telling management within these highly successful and recognized books?

The concepts and methodologies represented in this book are intended to give a thorough and complete understanding of the most effective principles behind the EMS audit process. Through proper execution and implementation, an organization may reap the benefits of a more visible understanding and commitment to:

- prevention of pollution;
- communication with relevant stakeholders;
- cost-effective continuous improvement;
- conformance to the designed EMS;

- a sound relationship with the environment;
- environmental regulatory compliance; and
- the organization's environmental policy, objectives, and targets.

The management organization may have as its ultimate goal the dedication to reaping these benefits. A clear and distinctive path to these goals is through the EMS and other related audit, assessment, verification, and surveillance methods carried out by the organization.

Key Definitions[4]

Before proceeding on, the author strongly recommends an extensive review of the key definitions of terms used in this book. Following many of the definitions are the referenced guides, standards, or organizations from which they originated. Following the definitions is a listing of the key acronyms commonly used in conjunction with the organizations, committees, and bodies, and terminology associated with the national and international efforts associated with the structuring, position, and development of the ISO 14000 series standards. Since we seem to live in a world of acronyms, it is also imperative that we understand and acknowledge these in order to better comprehend the concepts of this book and the standards themselves.

KEY DEFINITIONS

Accreditation. The procedure by which an authoritative body formally recognizes that a body or person is competent to carry out specific tasks. (ISO/IEC Guide 2)

Ancillary Material. Material input that is used by the unit process producing the product but is not used directly in the formation of the product. (ISO 14040)

Applicant. A legal entity applying for an environmental label for a product or a range of products, and which undertakes the compliance with ecological and product function criteria and the certification and costs involved in the application and awarding of the label. (ISO 14024)

Assessment. An estimate or determination of the significance, importance, or value of something. (ASQC Quality Auditing Technical Committee)

Assessment Body. A third party that assesses products and registers the quality systems of suppliers.

Assessment System. Procedural and managerial rules for conducting an assessment leading to issue of a certification document and its maintenance.

Audit. A planned, independent, documented assessment to determine whether agreed-upon requirements are being met. (ASQC Quality Auditing Technical Committee)

Audit Conclusion. Professional judgement or opinion expressed by an auditor about the subject matter of the audit, based on and limited to reasoning the auditor has applied to audit findings. (ISO 14010)

Audit Criteria. Policies, practices, procedures, or requirements against which the auditor compares collected audit evidence about the subject matter. (ISO 14010)

Audit Findings. The result of the evaluation of the collected audit evidence compared against the agreed audit criteria. (ISO 14010)

Audit Program. The organizational structure, commitment, and documented methods used to plan and perform audits. (ASQC Quality Auditing Technical Committee)

Audit Team. A group of auditors, or a single auditor, designated to perform a given audit. The audit team may also include technical experts and auditors in training. (ISO 14010)

Auditee. The organization to be audited. (ISO 14010)

Auditor (environmental). A person qualified to perform environmental audits. (ISO 14010)

Certification. A procedure by which a third party gives written assurance that a product, process, or service conforms to specified requirements. (ISO/IEC Guide 2)

Certified. The EMS of a company, location, or plant is certified for conformance with ISO 14001 after it has demonstrated such conformance through the audit process. When used to indicate EMS certification, it means the same thing as registration.

Client. The organization commissioning the audit. (ISO 14010)

Comparative Assertion. An environment claim made publicly regarding the superiority of one product versus a competing product which performs the same or similar function. (ISO 14040)

Compliance. An affirmative indication or judgement that the supplier of a product or service has met the requirements of the relevant specifications, contract, or regulation; also the state of meeting the requirements. (ANSI/ASQC A3) (See also conformance)

Conformance. An affirmative indication or judgement that a product or service has met the requirements of the relevant specifications, contract, or regulation; also the state of meeting the requirements. (ANSI/ASQC A3) (See also compliance)

Conformity Assessment. A conformity assessment includes all activities that are intended to assure the conformity of products or systems to a set of standards. This can include testing, inspection, certification, quality system assessment, and other activities.

Continual Improvement. The process of enhancing the environmental management system to achieve improvements in overall environmental performance in line with the organization's environmental policy. Note— The process need not take place in all areas of activity simultaneously. (ISO 14001)

Contractor. The organization that provides a product to the customer in a contractual situation. (ISO 8402, Clause 1.12)

Convention. A customary practice, rule, or method. (ASQC Quality Auditing Technical Committee)

Corrective Action. An action taken to eliminate the causes of an existing nonconformity, defect, or other undesirable situation in order to prevent recurrence. (ISO 8402, Clause 4.14)

Customer. The ultimate consumer, user, client, beneficiary, or second party. (ISO 9004-3, Clause 3.4)

Design Review. A formal, documented, comprehensive, and systematic examination of a design to evaluate the design requirements and the capability of the design to meet these requirements and to identify problems and propose solutions. (ISO 8402, Clause 3.13)

Elementary Flow. Any flow of raw material entering the system being studied which has been drawn from the environment without previous human transformation; any flow of material leaving the system being studied which is discarded into the environment without subsequent human transformation. (ISO 14040)

EN 45000. A series of standards set up by the EC to regulate and harmonize certification, accreditation, and testing activities. Guides for assessment and accreditation of certification bodies and guides for bodies operating product certification systems are expected to be published by the end of 1995.

Environment. Surroundings in which an organization operates, including air, water, land, natural resources, flora, fauna, humans, and their interrelation. Note—surroundings in this context extend from within an organization to the global system. (ISO 14001)

Environmental Aspect. An element of an organization's activities, products, and services that can interact with the environment. (ISO 14001)

Environmental Audit. A systematic, documented verification process of objectively obtaining and evaluating audit evidence to determine whether specified environmental activities, events, conditions, management systems, or information about these matters conform with audit criteria, and the communication of the results of this process to the client. (ISO 14001)

Environmental Impact. Any change to the environment, whether adverse or beneficial, wholly or partially resulting from an organization's activities, products, or services. (ISO 14001)

Environmental Labeling-Type 1. Multiple criteria-based, third-party voluntary environmental labeling program. (ISO 14024)

Environmental Management System (EMS). Organizational structure, responsibilities, practices, procedures, processes, and resources for developing, implementing, achieving, reviewing, and maintaining the environmental policy. (ISO 14001)

EMS Audit. A systematic and documented verification process to objectively obtain and evaluate evidence to determine whether an organization's environmental management system conforms to the EMS audit criteria set by the organization and to communicate the results of this process to management. (ISO 14001)

EMS Audit Criteria. Policies, practices, procedures, or requirements such as those covered by ISO 14001 and, if applicable, any additional EMS requirements against which the auditor compares collected evidence about the organization's EMS. (ISO 14011)

Environmental Objective. The overall environmental goal, arising from the environmental policy, that an organization sets itself to achieve, and which is quantified where practicable. (ISO 14001)

Environmental Performance. The measurable results of the environmental management system, related to an organization's control of its environmental aspects, based on its environmental policy, objectives, and targets.

Environmental Policy. A statement by the organization of its intentions and principles in relation to its overall environmental performance, which provides a framework for action and for the setting of its environmental objectives and targets. (ISO 14001)

Environmental Target. A detailed performance requirement, quantified wherever practicable, applicable to the organization, or parts thereof, that arises from the environmental objectives and that needs to be set and met in order to achieve those objectives. (ISO 14001)

Finding. A conclusion of importance based on observation(s). (ASQC Quality Auditing Technical Committee)

Follow-up Audit. An audit whose purpose and scope are limited to verifying that corrective action has been accomplished as scheduled and to determining that the action effectively prevented recurrence. (ASQC Quality Auditing Technical Committee)

Functional Unit. A measure of performance of the main functional output of the product system. (ISO 14040)

LCA Impact Assessment. The phase of life-cycle assessment aimed at understanding and evaluating the magnitude and significance of environmental impacts based on the life-cycle inventory analysis. (ISO 14040)

Inspection. Activities such as measuring, examining, testing, and gauging one or more characteristics of a product or service and comparing these with specified requirements to determine conformity. (ISO 8402, Clause 3.14)

Interested Party. An individual or group concerned with or affected by the environmental performance of an organization. (ISO 14001)

Joint Assessment. Cooperative assessments resulting in formal mutual recognition of certifications.

Lead Auditor (environmental). A person qualified to manage and perform environmental audits. (ISO 14010)

LCA Inventory Analysis. The phase of life-cycle assessment involving compilation and quantification of inputs and outputs for a given product system throughout its life cycle. (ISO 14040)

Life-cycle Assessment (LCA). Compilation and evaluation, according to a systematic set of procedures, of the inputs and outputs of materials and energy and the associated environmental impacts directly attributable to the function of a product throughout its life cycle. (ISO 14040)

Nonconformity. The nonfulfillment of a specified requirement. (ISO 8402, Clause 3.20)

Notified Body. A notified body is a testing organization that has been selected to perform assessment activities for a particular directive(s). It is approved by the competent authority of its member state and notified to the European Commission and all other member states.

Organization. A company, corporation, firm, enterprise, or institution, or part or combination thereof, whether incorporated or not, public or private, that has its own function and administration. (ISO 14001)

Organizational Structure. The responsibilities, authorities, and relationships, arranged in a pattern, through which an organization performs its functions. (ISO 8402:1994)

Practitioner. Private or government-sponsored entity, which develops, owns, sponsors, and/or administers an environmental labeling program. (ISO 14024)

Prevention of Pollution. Use of processes, practices, materials, or products that avoid, reduce, or control pollution, which may include recycling, treatment, process changes, control mechanisms, efficient use of resources, and materials substitutions. (ISO 14001)

Procedure. A specified way to perform an activity. (ISO 8402:1994)

Process. A set of interrelated resources and activities which transform inputs into outputs. (ISO 8402:1994)

Product. Goods and services for consumer, commercial, and industrial purposes. (ISO 14024)

Product Criteria. Set of qualitative and quantitative technical requirements that the applicant, product, or product category shall meet to be awarded an environmental label. Product criteria include ecological and product function elements. (ISO 14024)

Protocol Agreement. An agreement signed between two organizations that operate in different but complementary fields of activity and that commit themselves to take into account their respective assessment results according to conditions specified in advance.

Purchaser. The customer in a contractual situation. (ISO 8402:1994)

Qualification Process. The process demonstrating whether an entity is capable of fulfilling specified requirements. (ISO 8402:1994)

Raw Material. Primary or secondary recovered or recycled material that is used in a system to produce a product. (ISO 14040)

Registration. Procedure by which a body indicates relevant characteristics of a product, process, or service, or particulars of a body or person, and then includes or registers the product, process, or service in an appropriate publicly available list. (ISO/IEC Guide 2)

Requirements of Society. Requirements including laws, statutes, rules and regulations, codes, environmental considerations, health and safety factors, and conservation of energy and materials. (DIS 9004-3, Clause 3.3)

Responsible Care®. Comprehensive guidelines for environmental management systems adopted by Chemical Manufacturers Association (CMA) in 1988. Participation by individual businesses is an obligation of membership in the CMA.

Root Cause. A fundamental deficiency that results in a nonconformance and must be corrected to prevent recurrence of the same or similar nonconformance. (ASQC Quality Auditing Technical Committee)

Service. The result generated by activities at the interface between the supplier and the customer and by supplier internal activities to meet the customer needs. (ISO 8402:1994)

Service Delivery. Those supplier activities necessary to provide the service. (ISO 8402:1994)

Specification. The document that prescribes the requirements with which the product or service must conform. (ANSI/ASQC A3)

Stakeholders. Those groups and organizations having an interest or stake in a company's EMS program (e.g., regulators, shareholders, customers, suppliers, special interest groups, residents, competitors, investors, bankers, media, lawyers, insurance companies, trade groups, unions, ecosystems, cultural heritage, and geology).

Subcontractor. An organization that provides a product to the supplier. (ISO 8402:1994)

Supplier. An organization that provides a product to the customer. (ISO 8402:1994)

Survey. An examination for some specific purpose; to inspect or consider carefully; to review in detail. (ASQC Quality Auditing Technical Committee)

System Boundary. Interface between the product system being studied and its environment or other systems. (ISO 14040)

Technical Expert. Person who provides a specific knowledge or expertise to the audit team, but who does not participate as an auditor. (ISO 14010)

Testing. A means of determining an item's capability to meet specified requirements by subjecting them to a set of physical, chemical, environmental, or operating actions and conditions. (ANSI/ASQC A3)

Third Party. Person or legal entity recognized as being independent of the parties involved in the sale of a product. The practitioner, or its agent, is a third-party. Suppliers or producers are the first party, and consumers the second party. (ISO 14024)

Verification. Process of authenticating evidence. (ISO 14010) The act of reviewing, inspecting, testing, checking, auditing, or otherwise establishing and documenting whether items, processes, services, or documents conform to specified requirements. (ANSI/ASQC A3)

Waste. Any output from the product system which is disposed of. (ISO 14040)

KEY ACRONYMS[5]

Organizations

ANSI	American National Standards Institute. Adopts but does not write American standards. Assures that member organizations that do write standards follow rule of consensus and broad participation by interested parties. ANSI is the U.S. member of ISO.
ASQC	American Society for Quality Control, an administrator of the U.S. TAG to TC 207.
ASTM	American Society for Testing and Materials, a U.S. standards-making body that has developed a standard on EMS evaluation. It is an administrator of the U.S. TAG to TC 207.
CEN	Comité Européen de Normalisation (European Committee for Standardization)

CENELEC	Comité Européen de Normalisation Électrotechnique (European Committee for Electrotechnical Standardization)
CMA	Chemical Manufacturers Association
CSA	Canadian Standards Association
DoD/DOE	U.S. Department of Defense, U.S. Department of Energy
EAF	Environmental Auditing Forum
EAR	Environmental Auditing Roundtable
EC	European Commission
EEC	The European Economic Community. This comprises the EC and EFTA countries.
EFTA	European Free Trade Association. A group of nations whose goal is to remove import duties, quotas, and other obstacles to trade and to uphold nondiscriminatory practices in world trade. Members as of May 1995 were Austria, Finland, Iceland, Norway, Sweden, and Switzerland.
EPA	U.S. Environmental Protection Agency
EU	European Union. The EU is a framework within which member states have agreed to integrate their economies and eventually form a political union. Current members are Austria, Belgium, Denmark, Finland, France, Germany, Greece, Ireland, Italy, Luxembourg, Netherlands, Portugal, Spain, Sweden, and the United Kingdom.
GEMI	Global Environmental Management Initiative. A Washington, D.C.-based organization that deals with international business and environmental issues.
IAAR	Independent Association of Accredited Registrars
ICC	International Chamber of Commerce. An organization based in Paris, France.
ICSP	Inter-agency Committee on Standards Policy (within NIST)
IEA	Institute of Environmental Auditing
IEC	International Electrotechnical Commission, a parallel standards body to ISO in the electrochemical field.
IQA	Institute for Quality Assurance
ISO	International Organization for Standardization. A worldwide federation of national standards bodies (111 as of September 1995). ISO produces standards in all fields, except electrical and electronic (which are covered by IEC). Formed in 1947.
NACCB	National Accreditation Council for Certification Bodies (UK) now privatized as the United Kingdom Accreditation Service (UKAS). The British authority for recognizing the competence and reliability of organizations that perform third-party

	certification of products and registration of quality systems. Formed in 1984.
NCSCI	National Center for Standards and Certification Information (within NIST)
NIST	National Institute of Standards and Technology, U.S. Department of Commerce
NSFI	NSF International, a U.S. standards-making body that developed NSF 110 EMS standard, similar to ISO 14001.
NTIS	U.S. National Technical Information Service
OSHA	U.S. Occupational Safety and Health Administration
RAB	Registrar Accreditation Board. A U.S. organization whose mission is to recognize the competence and reliability of registrars of quality systems and to achieve international recognition of registrations issued by accredited registrars. It is seeking to be the U.S.'s ISO 14001 accreditation body.
RvC	Raad voor de Certificatie (Dutch Council for Certification). The Dutch authority for recognizing the competence and reliability of organizations that perform third-party certification of products, accreditation of laboratories, and registration of quality and environmental systems. The first such organization, formed in 1980.
SAGE	Strategic Advisory Group on Environment, formed by ISO to assess the need for an international EMS standard—since disbanded.
SCRAG	Standards Conformance Registrar Advisory Group, formed by the U.S. TAG to develop criteria for a U.S. accreditation body—since disbanded.
UKAS	United Kingdom Accreditation Service. (See NACCB)
WTO	World Trade Organization

General

CD	Committee Draft
DFE	Design for the Environment
DIS	Draft International Standard
EA	Environmental Audit
EAC	European Accreditation Certification
EAPS	Environmental Aspects in Product Standards
EL	Environmental Labeling
EMAR	Eco-Management and Audit Regulation (EU)
EMAS	Eco-Management and Audit Scheme (EU)
EMS	Environmental Management System

EPE	Environmental Performance Evaluation
EPI	Environmental Performance Indicators
GATT	General Agreement on Tariffs and Trade
IESU	International Environmental Systems Update
LCA	Life-Cycle Assessment (for Life-Cycle Analysis)
MRA	Mutual Recognition Agreement
NAFTA	North American Free Trade Agreement
SC	Subcommittee
SIC	Standard Industrial Classification code
SME	Small and Medium-Sized Enterprises
ST	SubTAG, Subgroup of TAG
SubTAG	Subgroup of TAG
SWG	Subworking Group (U.S. TAG)
TAG	Technical Advisory Group. A team used in the United States for groups that are responsible for input on international standards within their respective scopes; other countries may use other terms.
TC	Technical Committee, as in TC 207
TG	Task Group
WD	Working Draft
WG	Working Group

Training the Internal EMS Auditor

Training audit personnel in the techniques and methodologies of environmental sciences, performance, and management systems requirements is both necessary and profitable. This is really no different than the preparedness required for any highly specialized skill. However, the vague assumption is often made that the qualification criteria for becoming an EMS auditor is automatically satisfied by merely working in the environmental field or being a previously qualified QSA. The ability to perform rapid, cost-effective EMS audits requires unique traits and techniques acquired through formal, on-the-job *training and experience.*

What training can be given to the individual who appears to be suited for the EMS internal audit function? For example, the author has inquired to several of the existing accreditation bodies for information regarding internal EMS auditor course accreditation. As of February 1997, there are no formal systems in place to "certify" internal EMS auditors. Further, the existing national and international bodies have agreed that the level of training necessary at this time for the internal auditor is a 3-day supplemental EMS auditor course (criteria currently under development) or a 36–40-hour accredited

EMS auditors course. Of course, there is no requirement for EMS internal auditors to be "certified." Accordingly, the author strongly advises management to consider specialized in-house internal audit courses "tailored to your unique requirements." For small and medium-sized enterprises (SMEs), the organization should consider use of an accredited EMS audit course provider.

Which brings us to the issue of selection of a qualified training organization.

One of the biggest concerns in the training industry today is the *experience* of that organization in the environmental audit profession. Many of the "gurus" of quality are, of course, jumping on the ISO 14000 bandwagon as if they were training environmental auditors for decades. Be selective. Look for accredited organizations. Require instructor qualifications. Ensure that these individuals have ample experience in the environmental compliance and EMS audit training professions. If formal training is not immediately available and management desires to put the new EMS auditor to work, as a minimum:

(1) Have the auditor work with an experience and qualified individual for several in-plant or on-site environmental audits.

(2) Have them read this book cover to cover *more than once*.

Afterward, the trainee can be carefully initiated into the EMS internal audit program, which can be an effective training device providing immediate benefits derived from the audit review.

After a complete understanding of the techniques for EMS audits has been developed, it has been found to be effective to advance to the on-the-job training method. EMS audit management may discreetly monitor the performance of the trainee. Strengths and weaknesses are discussed in detail. A comprehensive review of the auditor's next assignment can ensure that the appropriate corrective actions were taken throughout the audit process. Each audit thereafter is carefully monitored, while the trainee's area of responsibility may be enlarged. The length of time to ensure independent audit responsibility depends, of course, on the capabilities of the auditor.

Remember, formal classroom training, which can take up to five days, can reduce the learning time by six months to one year. Without adequate guidance, a new EMS auditor will make errors, false starts, use poor techniques, do too much, not enough, talk too much, not ask the right questions, and/or fall into the many pitfalls that take years to recognize. Training can reduce much of the aggravation.

Auditor Characteristics

Recognition that not all individuals will make effective EMS auditors is critical. Experience has shown that of those who attend formal auditor training, 15–20% are ill-suited for the demands which are inherent on the auditors.

This is not alarming or unusual. Not everyone is programed to perform every specific task. The problem is that management does not always recognize the personal traits, knowledge, and attitudes which are required.

The following positive auditor characteristics have been developed over a long period of time to best relate what it takes to be an effective auditor.

DESIRABLE AUDITOR CHARACTERISTICS

- Knowledge of requirements
- Knowledge of techniques
- Sound judgement
- Patience
- Interest
- Tenacity (strong)
- Professional attitude/integrity
- Good listener
- Unafraid of being the "bad person"
- Inquisitive
- Communicative at all levels
- Analytical
- Honest
- Diplomatic
- Disciplined
- Good planner

Knowledge Traits

Knowledge of the five basic disciplines—environmental sciences, technology, legislation, quality and environmental principles and practices—is acquired by extensive work in the industrial environment or by considerable formal training. Some auditors find themselves auditing a discipline that they know little about.

Specifications and requirements are often taken as obvious because the auditor has read them once or twice or they have been around for a long time. Experience has shown that the auditor requires closer involvement with these documents as to interpretation and how they are implemented. There is very little substitution for working with the requirements in the learning process. Evaluation techniques are acquired only through formal training, good on-the-job indoctrination, or much self-study and experience. Individuals will not automatically go out and perform comprehensive and adequate evaluations without assistance. It is difficult to perceive how an auditor can perform an audit without a thorough understanding of the knowledge characteristics shown. Surprisingly, some individuals attempt to perform audits with little background by way of experience and education. This background can be gained by practical experience or well-designed class studies in formal training.

Inherent Personality Traits

To make sound judgements, the auditor must always keep an open mind on all observations being made. The auditor must gather the facts, weigh them against personal experience and knowledge of the requirements, and then make a decision regarding the adequacy of whatever is being investigated, be it instructions, procedures, processes, tests, programs, materials, equipment, or whatever. Inherently, some people have great difficulty making decisions; if so, they will have considerable frustrations operating in the auditor's world, for the auditor must make judgements and decisions as a routine function of the job.

Occasionally, an auditor may go into a department or facility with preconceived opinions about the program or the overall conditions. When this occurs, the chances of the auditor making an objective evaluation are reduced considerably.

The auditor should have studied all the information obtainable concerning the requirements, procedures, and product during preparations for the audit. But, it is important to remember that the information gathered is background information to be used in making the actual investigation. On-site, the auditor must maintain a "show me" attitude and assume nothing. The auditor must (1) have the patience to perform the investigations necessary to ensure that the details of the requirements are, in fact, being adhered to; (2) the tenacity to stay with a situation regardless of how much bother it may be to unearth the facts and arrive at sound, adequate judgements; and (3) possess the internal strength to stick to the arrived-at conclusions.

The auditor does not draw firm conclusions until the operation in question has been thoroughly investigated. The auditor may find that the errors are closer to home than was previously realized. For example, armed with the fact that the activity had released material that did not meet certain environmental test specifications, the auditor began preparing a corrective action report. Further investigation revealed that the revision to the environmental test specification had not been imposed and the activity could not comply with a requirement that had never been received. Actually, there were two auditing errors. The auditor had not checked to determine if all documentation had been submitted and had prejudged the activity's performance.

To repeat, the auditor must maintain an open mind in performing any investigation, verification, or confirmation. One of the worst things that can happen is for an auditor to whitewash an inadequate EMS. Some management will not implement needed improvements if the activity appears to be compliant. Therefore, auditors who do not report significant deficiencies can cause real harm to the activity's personnel, as well as to the auditor's management.

Performing audits is hard mental and physical work; therefore, if auditing is not interesting to the individual, that person probably will not be an effective auditor.

A professional attitude with integrity is an must for the auditor. It must be remembered that the auditor represents the activity in all personal actions; thus, the auditor's personal competency also reflects on the activity.

There is a direct relationship with the manner in which the auditor and personnel being audited interact. The way auditors present themselves is directly proportional to the way they are treated. A positive professional approach should be received in kind, while unprofessionalism will be treated accordingly.

Unafraid of being the "bad person" means that the facts that are uncovered may reflect on someone's performance. This may make the auditor very unpopular. Since the auditors must "call it as it is," they cannot concern themselves with personalities. The auditor investigates systems but does not attack people. It is the audited organization's management's responsibility to ensure that the system is operating in accordance with requirements.

Inquisitiveness and analytical tendencies are natural traits that will help the auditor with the investigation. If finding out things is interesting, the auditor will be inquisitive. If the auditor has the ability to analyze facts and sort out opinions, analytical talents will be evident in that person.

Honesty is a must. The auditor has nothing to hide. Everything is laid out, "as it is." There are no "I got ya's" sprung on the activity at the closing meeting.

Diplomacy is not easily acquired. Some people have it, and others must work very hard at it. The ability to tell people that they are doing very poor jobs, and making them like it, is rare. For the auditor, attention should be kept on the system of control, while the deficiencies that are observed can be stressed as potential areas of improvement in the program. The disciplined auditor will keep everything on track, knowing what must be accomplished and how to get there. The auditor allows no outside influence to interfere with the objectives of the audit.

It is not certain if being a good planner comes through training or if some individuals are naturally more organized than others. Be that as it may, auditor planning is critical in effective performance. Planning before the audit will prepare the auditor for the verifications that must be made. Planning is the backbone of effective auditing.

Undesirable Auditor Characteristics

Of course, the following undesirable characteristics are the opposite of the desirable; however, it is worthwhile to also examine these:

UNDESIRABLE AUDITOR CHARACTERISTICS

- Argumentativeness
- Opinionated
- Lazy/lack of desire
- Easy to influence
- Inflexible
- Jumps to conclusions
- Mr./Ms. "Nice Person"
- Believes everything he/she hears
- Noncommunicative
- Devious
- Poor planner
- Unprofessional

Arguing has no place in the audit process. Let us define arguing as" heated discussion that becomes emotional." Nothing is accomplished through arguing except frustration. The auditor can agree to disagree. What is important is that each position is understood. Occasional findings, deficiencies, and conclusions may be brought to higher authorities for resolution. The auditor can accept this, but will not be drawn into arguments.

People have opinions on nearly all subjects. When the opinions of the auditor interfere with the ability to perform the tasks at hand, the auditor will be ineffective. Some individuals have an attitude of "don't confuse me with the facts, my mind is already made up!" Auditors must use their experiences, background, and common sense to make sound and just conclusions.

Communication by the auditor is how information is gathered; therefore, the person who has difficulty expressing ideas and talking with other people is going to have considerable problems in auditing.

The poor planner "can't get there from here." The inability to plan will cause the auditor not to be prepared to perform the audit. During the audit, the auditor will try to do much of the work that should have been accomplished before the evaluation.

Unprofessional and arrogant individuals have no place dealing with outside activities. More damage will be done than good received. Such individuals should be screened out of any audit program.

One of the most difficult phases of audits is the auditor's "good person, bad person" syndrome. It has been said that a good auditor, like a good inspector, "ain't got no friends." Image is a serious matter with many people and, in criticizing another person's performance, auditors risk personal acceptance. This apparent need for acceptance definitely affects the auditor's ability to make sound judgements. This type of auditor is referred to as "the nice person."

An example of this occurred while the author was on an audit of a new activity. The team had uncovered many deficiencies in the operating functions, but the chairperson was the type who avoided hurting people's feelings. Consequently, when the closing meeting was presented, the chairperson reduced the findings to only insignificant deficiencies. In order to spare the activity's personnel (and the chairperson's), the chairperson was playing down some very serious problems.

The auditors are not out to deliberately aggravate anyone, but it is their function to call it like it is. This, of course, can make people unhappy because they feel that the deficiencies found are a direct reflection on their work. On the other side of the coin, many a quality manager has confided that it was desirable to have a given condition brought to light because previously, higher management ignored the problem and having it reported as a serious condition was the only way to get management support. In this instance, a corrective action can work to the quality manager's benefit. At the same time, auditors must be very careful that they are not being used by the quality manager or by another manager to gain some ulterior advantage not wholly related to the program.

Unfortunately, some individuals believe everything they hear and are extremely easy to influence. Since, in our daily lives, we generally accept things we hear for what they are worth, the auditor must remember to *verify* all verbal information. Occasionally, the auditor will be challenged by great indignation as to the findings. The auditor cannot be backed off by browbeating tactics nor be persuaded by sweet talk and sales pitches that are intended to influence rather that present new facts.

The auditor must be careful of making decisions before having all of the facts. Jumping to conclusions can cause embarrassment later.

Types of Auditors

Auditors are human beings. We carry with us many different types of personalities, as discussed on previous pages, which may positively or adversely affect the outcome of an audit. The following are various types of auditors that have surfaced over the years, followed by a detailed discussion of the impact these auditors may have on the overall audit processes:

TYPES OF AUDITORS

- Desk
- Cocktail
- What can you do for me?
- Sticker
- Know-it-all
- Nit-picker
- Out of the auditor's field or ad hoc
- The professional

The *desk* auditor is very concerned with the processes of ensuring all the documents (i.e., procedures, instructions, etc.) are in compliance with their applicable standard(s), specification(s), and/or regulation(s). They may spend too much time verifying that all of the "i's" are dotted and "t's" are crossed, but leave little time to actually verify that the activity is effectively implementing the system requirements. The auditor should carefully plan all phases of the audit, including the time frames and schedules developed to ensure that there is ample time to do both.

We are very happy to see over the years that the *cocktail* or *what can you do for me auditor* has been steadily decreasing. What's in it for you is the ability to effectively act as the eyes and ears of management in verifying, validating, and reporting the status of systems conformance. The key role in an organization's effort in determining a level of continuous improvement to be applied to all areas lies within the hands of an effectively managed audit and assessment process.

There are generally two different types of *stickers*. The first is the type who has *got to find something*. They feel concerned, and even intimidated, that if they do not have any adverse conditions to report, the organization's audit management will feel that they have not done their job. Additionally, they may tend to "stick" to one particular type of requirement until something is found. This may lead to the auditor getting off track due to the unnecessary time spent in one area.

The second type of *sticker*, after discussing several fairly undramatic conditions, may get lazy. They realize now that they have several items on which to report, which will look like they have done a complete job, but then spend little effort continuing through the duration of the audit process.

Any book on management systems auditing would be incomplete without discussing the world of the *nitpick* auditor. The nitpicker will make a very big deal out of something relatively small. This often occurs when the design of an organization's audit program does not contain clear definitions categorizing the importance of criticality of system nonconformances. Chapter 12 on EMS audit reporting will discuss in detail the methods for determining and prioritizing system nonconformances.

Ladies and gentlemen, there is no such thing as a *know-it-all* auditor. This personality trait appears to surface on several occasions, the first of which is when an auditor really does not know all of the specific technical and operational requirements of a given process, but certainly does not want the auditee to know that. They may shake their heads, "Oh, yes, yes, I understand," when what they are really thinking is, "Help!" Systems auditing is a learning process. "I am not entirely familiar with this process. Could you please walk me through and show me the various controlled methods applied to this operation?" is an appropriate response to uncertainty. This is especially important for internal auditors in effectively acquiring the technical and operational understanding of a given operation to better expand their abilities in effectively auditing this area.

There are also some auditors who, unfortunately, really think that they *do* know it all. This is quite common among system auditors who become very familiar with auditing to a specific requirement of a given management system standard. They become convinced that there is only one possible way these requirements can be effectively implemented. Second-party auditors (who may be customers or state or federal regulators having no direct responsibility for the organization's management systems) need to keep an open mind when reviewing operational conformance. There are, in fact, many different ways a facility may comply with the same set of generic management system requirements. The "show me" approach is extremely important in eliminating any level of bias regarding how an operation must be doing something.

The *professional*, again, is something we strive to be in our everyday work life. How one treats someone is a direct reflection on how one probably gets treated back. The author is encouraged to see the coordination and cooperative levels between the auditor and auditee grow in leaps and bounds over the years. This is not an "us against you" thing. Much discussion will be represented throughout the following chapters on breaking down the barriers and allowing a true professional approach and relationship to develop among all involved in the audit process.

EMS Audit Objectives

The EMS audit has as its primary objective the ability to bring to the forefront the specific data and information collected to be used in determining the appropriate corrective, preventive, and continuous improvement actions specific to the organization's management system requirements. Let us now explore some of these primary objectives:

PRIMARY OBJECTIVES OF AN EMS AUDIT

- Determination of the extent to which established policies and procedures conform to a given set of requirements.
- Determination of the extent of effective implementation and conformance to established policies, procedures, and instructions.
- Determination of the need for new procedures and instructions, or changes to existing documents.
- Assessment of the familiarity of management, supervision, and operational personnel with documentation to which they are required to perform.
- Determination of the need for requalification or additional training for organization personnel having specific responsibilities and authorities within the EMS.
- Reinspection or retesting, as necessary, of collective samples.
- Assurance of availability, use, and adherence to work, inspection, and/or monitoring criteria.
- Determination of effective sampling, data collection, and analytical processes.

Much has already been said in the world of systems auditing about "saying what you do and doing what you say." A key factor in this process is how well a facility's personnel really know their everyday work life. A comprehensive assessment of this knowledge may lead to the need for additional training to ensure effective implementation of given procedures and instructions.

An EMS auditor will, of course, verify and assess various analytical, environmental inspection and testing operations conducted at that facility. Our ability to witness the actual performance of personnel and systems is critical in determining the effectiveness of a facility's operations. Things that should be sought out by the auditor include, "How readily available are all of the applicable procedures, standards, specifications, and/or regulations?" Additionally, "How effectively are the sampling, data collection, and analytical processes carried out to complement the inspection and testing methods required?"

These key audit objectives will be explored in great detail throughout the appropriate chapters of this book. Determining at what stage these key objectives will surface, including how, when, where, and why, will become critical for the EMS auditor to consider in planning, preparing for, and executing an effective EMS audit.

Key Differences—QMS/EMS Auditors[6]

Ronald Black, Former Director of Environmental Health and Safety at BF Goodrich and past president of the Environmental Auditing Roundtable, says the degree to which quality management system auditors will have to fulfill additional requirements for EMS auditing will depend largely upon the interpretation of ISO 14001 in the marketplace.

He explains that the ISO 14012 guidelines on EMS auditor qualifications were developed in parallel with the development of the ISO 14001 EMS specification and ISO 14004 EMS guidelines. Many international experts participated in both efforts, and drafts were freely shared. TC 207 leadership monitored the development of drafts to further ensure that the EMS specification, guidance document, and audit guidelines were synergistic.

Thomas P. Ambrose, President of Health, Safety, and Environmental (HSE) Management, says many experts will view the question as posed, but he suggests it be framed in the broader context of what level of expertise is needed by any individual to perform competent conformance audits to the ISO 14001 EMS standard.

Ronald Black[7]

The guidelines for auditor certification can reasonably be thought to be those intended to qualify an auditor for EMS registration purposes.

ISO 9000 series standards on quality auditor qualifications are more general than those of ISO 14012. The QMS auditor guidelines more closely parallel general auditing qualifications of most auditing disciplines. The ISO 14012 guidelines, on the other hand, recommend specific experience in the following areas:

- *environmental science and technology;*
- *technical and environmental aspects of facility operations;*
- *relevant requirements of environmental laws, regulations, and related documents;*
- *environmental management systems and standards; and*
- *audit procedures, processes, and techniques.*

All other qualifications in 14012 are either generic to other audit disciplines, e.g., environment, health, and safety (EH&S), financial, or QMS, or can be acquired within a reasonable period of time. Therefore, an obvious response to the question of "What additional qualifications will a QMS auditor need to effectively audit EMSs?" is that auditors will need to acquire those skills identified in the 14012 guidelines....

Interpretation Comes First

However, the extent to which these skills are necessary to audit a facility for registration will depend on how the standard is interpreted.

If, like the QMS registration requirements, the systems need only be in place and conform with the ISO 14001 requirements, then most auditors having general skills such as interviewing and document review should qualify, especially if supported by the type of training QMS auditors are required to take. Under these conditions, most auditors, e.g., financial, environmental, or QMS, will qualify.

However, many ISO 14000 stakeholders believe that, to qualify for certification, facilities will be required to have more than a system in place and to meet the basic tenets of the standard. For example, one higher level of testing would require a determination as to whether the EMS is designed to effectively manage environmental aspects appropriate to the size and activities of the operation.

A second level of analysis would address the impact of these aspects on the environment and the ability of the systems to classify their significance. Both of these investigative extensions would require significant environmental training and experience. In addition, ISO 14001 requires a system component that ensures a commitment to legal compliance, but environmental regulations in the United States are extensive and complex.

How much regulatory experience, knowledge, and training does an auditor need to have to evaluate a system to ensure that it is capable of sustaining compliance with laws and regulations?

In conclusion, the additional qualifications needed by QMS auditors who do not have environmental experience and training will depend on what the certification body accreditors establish as registration requirements.

If the requirements are limited to systems in place, then little additional experience or training will be necessary, and auditors in most fields will qualify with little additional training. If the systems need to be evaluated for suitability and effectiveness, then a strong background in environment will [be] necessary.

Thomas P. Ambrose[8]

In my view, auditing the quality of an EMS will require individuals who have acquired a body of knowledge from working in the field over many years. They will have firsthand experience in business systems, operations, technology, quality, and environmental management.

In the United States, the process is under way to synchronize the adoption of ISO 14001 as the American national EMS standard, as well as the ISO-related EMS auditing guidelines.

The question of EMS auditor qualifications has received attention since the Strategic Advisory Group on Environment (SAGE) recommended moving forward to develop the ISO 14000 series in the early 1990s. ISO partially answered the question with the ISO 14012 auditor qualification standard, but almost everyone would agree that the qualifications outlined in this guideline are baseline and represent a floor for a country such as the United States, which has mature, sophisticated environmental management controls in place.

A national program will be established in the United States in 1996 to accredit ISO 14001 certification bodies, EMS registrars, auditor certifiers, and training providers.... The spotlight will be focused sharply during 1996 by the accreditation body on the key issue of qualifications of individuals who perform EMS conformance audits to ISO 14001.

QMS/EMS—Sharing a Common Basis

Detailing EMS auditor qualifications is too complex to be described in a few short paragraphs. However, an effective EMS strives to integrate environmental concerns into other routine business management processes.

For both quality and environmental management systems, effective integration and implementation occur when all employees are brought into the process to share a common objective and their efforts are linked with achieving the organization's strategic business goals. However, the management mind set in establishing implementation processes differs in the following important aspects:

- *Quality* is a contract between a supplier and customer (e.g., "producer" and "buyer"), which places the ultimate focus on increased buyer satisfaction. The quality process integrates fundamental management techniques in an organized, disciplined system focusing on continuous improvement. The goal is to connect everyone's daily effort in the producers' organization with the understanding of what the buyer wants to deliver a product or service that consistently meets or exceeds those expectations.

- *Environment* has stakeholders beyond the producer-buyer relationship who have legitimate interest in the process and influence over the outcome. In addition to meeting buyer expectations, the focus of the producer is to comply with government requirements, minimize environmental impacts, and address community concerns. New regulations and society's acceptance of the producer's operations and products are additional forces that drive the EMS's continuous improvement cycle and "license to operate."

Environmental and quality management systems have a number of similar approaches for establishing and documenting procedures to demonstrate that the system is working and conforming to internal policy and established standards. But as integration becomes more successful, there is less likely a need for numerous procedures providing a ready "audit trail."

The auditor must have sufficient experience to recognize that the quality of the EMS in place is achieving the organization's stated objectives.

Insights for EMS Auditor Qualifications

Earlier this year, one of my clients, NSF International, asked for help in shaping assessment tools to be used by eighteen organizations participating in its EMS demonstration project. The organizations represent a variety of products and services and range from Fortune 100 corporations to small manufacturers to government agencies. They have come together to learn more about EMS self-assessment and implementation issues and are primarily using the ISO/DIS draft 14001 as the benchmark.

The idea was to provide basic guidance to the NSF project organizations on the skills, on the job experience, and on training that "expert bodies" thought were minimal requirements to credibly assess environmental management systems.

Key Considerations for EMS Audit Teams

The introduction of ISO/DIS 14001 states that an organization can assure interested third parties that an appropriate EMS is in place by demonstrating successful implementation of the standard. It also points out that the specified requirements for the EMS may be objectively audited for certification/registration purposes and/or self-declaration purposes.

The decision to adopt ISO 14001 as an organization's EMS is a marketplace issue. However, for either a third party or self-declaration to stand up to public scrutiny, the auditors chosen will have to demonstrate competence to do the job. Audit team members should, at a minimum:

- be selected from appropriate management backgrounds to ensure peer review;

- be multidisciplinary, representing a mix of expertise ranging across management systems to control technology, with direct experience with the relevant type of operation;

- have direct experience with the type of operations at hand;

- collectively have appropriate expertise, knowledge, and proficiency in auditing techniques (e.g., verification, observation, information analysis); and

- be unbiased and display due diligence—assessment findings should be based on factual information gathered during the assessment process using detailed but selective testing, inspection, and interviewing.

While ISO 14001 does not require the use of certified auditors, a credible certification system will provide the marketplace with "a predictable level of confidence" that an appropriate EMS is in place.

Endnotes

1. This section originally was published in the first edition of the *ISO 14000 Questions and Answers*, published by CEEM Information Services, 10521 Braddock Road, Fairfax, VA 22032, USA. Tel: +1-703-250-5900; fax: +1-703-250-4117; E-mail: <inquiry@ceem.com>; Website: <http; //www.tregistry.com/ceem/>. Reprinted with permission.
2. Ibid.
3. Ibid.
4. Ibid.
5. Ibid.
6. Ibid.
7. This article, "What New Qualifications Will a QMS Auditor Need?—Key Differences" (Ronald Black, Thomas P. Ambrose) originally was published in the January 1996 issue of the *International Environmental Systems Update* (IESU) published by CEEM Information Services, 10521 Braddock Road, Fairfax, VA 22032, USA. Tel: +1-703-250-5900; fax: +1-703-250-4117; E-mail: <inquiry@ceem.com>; Website: <http; //www.tregistry.com/ceem/>. Reprinted with permission.
8. Ibid.

2

The ISO 14000 Series Audit Standards

ISO 14010 Guidelines for Environmental Auditing: General Principles◆

The primary purpose of this guidance standard is to provide general principles of all types of environmental audits. The key word, as indicated in the title itself, is "general." The intent of the standard is to guide organizations, auditors and their clients through the general principles common to the execution of environmental audits but, quite frankly, it does not begin to address the gamut of multi-disciplined issues that face the environmental auditor on a daily basis.

Included in this standard are the scope, normative references, and key definitions pertinent to the principles addressed within the standard. For the purpose of a one-stop review, the author, as you have already read, included these definitions in Chapter 1.

For you, the reader, this overview will take a closer look at each element represented in the standards (14010, 14011, and 14012) and reference the specific chapters within this book that will greatly expand our understanding of the guidance represented in these standards.

4   Requirements for an Environmental Audit

An environmental audit should focus on clearly defined and documented subject matter....◆

Let us keep in mind that this guidance standard recommends, for your consideration, principles to effectively control and manage the environmental audit processes. Therefore, the word "should" can be interpreted as methods for improvement, guidance, and food for thought to enhance the execution and performance of the environmental audit processes. A clear definition of the various roles, responsibilities, and authorities of all parties involved in the audit is essential prior to commencing with the audit. This will best coordinate the communicative efforts and establish a working relationship to assure a common goal is achieved...conformance or compliance with specified requirements.

> *...the audit should only be undertaken if, after consultation with the client, it is the lead auditor's OPINION that...*◆

Hmm. Needless to say, there is some concern initially with a guidance standard that places such emphasis on opinions. The book in its entirety addresses the audit process being based on facts and conclusions where objective evidence has indicated the issues at hand are not just opinions of one individual involved in the audit process. Clear, concise objective evidence should be disclosed regarding the lead auditor's review to determine if:

- *there is sufficient and appropriate information about the subject matter of the audit;*
- *there are adequate resources to support the audit process; and,*
- *there is adequate cooperation by the auditee.*◆

5   General Principles

5.1   Objectives and Scope

> *The audit should be based on objectives defined by the client....*◆

The client? OK, fine. Let's say the client has defined the objectives for a registrar conducting a third-party audit for certification to ISO 14001. The client has indicated the objectives of the audit will only include those products, activities, and services conducted on-site. Ladies and gentlemen, the client does not set these objectives; the registrar does. This will assure all elements of the standard will be reviewed and evaluated accordingly to assure conformance with the standard for operations, activities, products, and services associated with the organization's system, whether or not all these activities are to be verified on-site or not. Through this effort between the audit function, auditee, and client, the scope and objectives of the audit may be determined by:

- *the audit program management, in consultation with the client, and*
- *the description of the extent and boundaries of the audit based on facts and conclusions of the review and audit preparation processes.*◆

This, in turn, will allow for a broader understanding of the objectives and scope to be communicated to the auditee, with their input, prior to the audit.

This point illustrates an example of the absence of true direction, concepts, methodologies, and techniques regarding "all types of environmental audits."

5.2   Objectivity, Independence, and Competence◆

This element of the general guidance standard does not address competence, or rather, the methods to be considered for assuring an appropriate combination of knowledge, skills, and experience to carry out audit responsibilities. The standard suggests the members of the audit team should be independent of the activities they audit.

Remember, this guidance standard is for all types of environmental audits...including internal EMS audits to ISO 14001, right?

Interesting to note here that ISO 14001, Section 4.5.4, Environmental management systems audits, does not require the auditor to be independent of the activities audited (see Chapter 5). Audits should certainly be carried out free from bias. Many of the concepts represented in this book give due consideration to a self-assessment process conducted by auditors having direct responsibilities relevant to the activities, products, and services for a more concentrated, comprehensive understanding, not only of the audit process itself, but especially for conformance to requirements these individuals are directly responsible for in the first place.

5.3   Due Professional Care◆

The care, diligence, skill, and judgement of an auditor relies entirely on the competence of management of the audit organization. Confidentialities and discretion are usually as advised by the client, auditee, and/or audit organization regarding the disclosure of information, findings, and/or documents obtained during the audit and included in any type of report submitted for review and response.

All types of environmental auditors should have the freedom to conduct the audit process in a skilled and professional manner that promotes the advancement and continuous improvement effort of the audit organization.

5.4   Systematic Procedures◆

Reference is made here to ISO 14011, Guidelines for conducting environmental management system audits. Not exactly applicable to all types of environmental audits. Additional reference to this topic is addressed in Chapters 4 and 5, respectively. Consideration should certainly be given to the procedures for conducting, for example, EMS internal audits and how effective these programme procedures are for other types of environmental audits.

5.5   Audit Criteria, Evidence, and Findings◆

Audit criteria are defined as policies, practices, procedures, or requirements against which the auditor compares collected audit evidence. In most cases, the audit criteria will be distinctly different for one type of environmental audit from another. Chapter 7, EMS Audit Planning, will go into greater detail with respect to the development and use of appropriate audit criteria.

Audit objective evidence is based on the facts and conclusions of the audit organization from collected, analyzed, verified, interpreted, and documented subject matter in an examination and evaluation process to determine the following:

- *whether or not audit criteria is met;*
- *whether or not conformance to specified requirements has been achieved; and,*
- *the scope, purpose, intent and objectives of the audit have been fulfilled (Chapter 10).*◆

5.6   Reliability of Audit Findings and Conclusions◆

This section, unfortunately, addressed the fact that auditors rarely have enough time to do everything they need to do in conducting a comprehensive, effective audit. By investigating only a sample of information available, there is an element of uncertainty inherent in all types of environmental audits. This uncertainty is at the core of what the audited organization should investigate to determine the extent and magnitude of nonconformances encountered relevant to the organization's EMS and/or environmental program require-ments. Chapters 10 through 12 will provide extensive detail regarding the reliability of audit findings, conclusions, and the final section of ISO 14010, Reporting.

The communications and coordination of all involved in the environmen-tal audit may assure increased visibility in the audit processes by a mutual understanding of the guidance given in ISO 14010. Auditors, auditees and clients alike should consider the fundamental guidance presented in this standard during initial indoctrination and awareness training to further pro-mote a clear understanding of the basic mechanics on a competent, reliable audit process.

ISO 14011 Guidelines for Environmental Auditing: Audit Procedures—Auditing of Environmental Management Systems◆

The scope and purpose of this guidance standard establishes audit proce-dures for the conduct of EMS audits through planning and performance of the audit to determine conformance with EMS audit criteria. Audit organizations should consider the use of this guidance standard during the development, implementation, and continuous improvement of a comprehensive EMS audit programme and procedures for the purpose of conducting ISO 14001 EMS audits.

4   EMS Audit Objectives, Roles, and Responsibilities
4.1   Audit Objectives

Clearly defined scope, purpose, intent, and objectives should be developed for each type of EMS audit to ISO 14001 performed. The following typical objectives should be "tailor-made" to suit the specific element, requirement, and/or program topic to be covered by the EMS auditor and audit process:

- *to determine conformance of an auditee's EMS against the EMS audit criteria;*
- *to determine whether the auditee's EMS has been properly implemented and maintained;*
- *to identify areas of potential improvement in the auditee's EMS;*
- *to assess the ability of the internal management review process to ensure the continuing suitability and effectiveness of the EMS;*
- *to evaluate the EMS of an organization where there is a desire to establish a contractual relationship, such as with a potential supplier or a joint venture partner.*◆

4.2   Roles, Responsibilities, and Activities
4.2.1   Lead Auditor◆

A comprehensive listing of lead auditor responsibilities and activities is represented. Chapters 7 through 12 will address considerable discussion, concepts, and techniques directly related to the guidance requirements of this section, including auditor (4.2.2) and team coordination and representation (4.2.3).

4.2.4   Client

This section addresses specific responsibilities and activities covered by the client. The client is defined as the organization commissioning the audit. The following is the listing within ISO 14011 of these suggested responsibilities:

- *determining the need for the audit;*
- *contacting the auditee to obtain its full cooperation and initiating the process;*
- *defining the objectives of the audit;*
- *selecting the lead auditor or auditing organization and, if appropriate, approving the composition of the audit team;*

- *providing appropriate authority and resources to conduct the audit;*
- *consulting with the lead auditor to determine the scope of the audit;*
- *approving the EMS audit criteria;*
- *approving the audit plan;*
- *receiving the audit report and determining its distribution.*◆

The author wishes to address several of these bullet issues for your consideration to allow for the most effective, comprehensive EMS audit possible.

selecting the lead auditor...approving the composition of the audit team....◆

Hmm. Let's think about this for a minute.

The client requests a registrar to conduct a third-party EMS audit for certification to ISO 14001. The client requests resumes and auditor qualifications of the registrar's auditors in order to choose who is going to do the audit.

Well, maybe the client should be responsible for this.... Good luck.

Of course, the intent of this guidance surely doesn't suggest the client will "mold" the registrar audit team to their specific requirements, right?

Oh, and by the way, don't call me "Shirley."

The client should, however, assure that the audit personnel, especially those from a registrar, have some relevant environmental experience within the industry of the client's facility being audited.

approving the EMS audit criteria....◆

Hmm. The author has been involved in several contracted environmental program audits where we were required to submit our audit plan, protocol, and procedures for conducting the audit. In each case, the client requested us to disregard entirely several items within the plan and procedures. Again, in each case, the client was informed that these were physical requirements within the standards or regulations that we were required to audit against. Didn't matter. In one incident, several of the "disregarded" requirements were audited by the EPA, who proceeded to "blow them off the map" with respect to a plethora of nonconformances against these requirements. Food for thought.

Qualified, professional auditors have a keen insight into the requirements and an effective method for evaluation of conformance to these requirements. The client should always take into consideration this insight for conducting the best, most thorough audit possible.

approving the audit plan....◆

Ditto previous comments.

By the way, who is the client regarding the most common type of EMS audit that will ever be performed, the internal EMS audit?

4.2.5 Auditee

The following guidance from ISO 14011 covers types of responsibilities and activities by the auditee:

- *informing employees about the objectives and scope of the audit as necessary;*
- *providing the facilities needed for the audit team in order to ensure an effective and efficient audit process;*
- *appointing responsible and competent staff to accompany members of the audit team, to act as guides to the site, and to ensure that auditors are aware of health, safety, and other appropriate requirements;*
- *providing access to the facilities, personnel, and relevant information and records as requested by the auditors;*
- *cooperating with the auditors to permit the audit objectives to be achieved; and*
- *receiving a copy of the audit report unless specifically excluded by the client.*◆

Good guidance here. The author strongly recommends all management level personnel responsible for the organization's EMS to develop an "Auditee plan" for any type of audit conducted at the facility. Use this as auditable criteria to verify and validate that these key objectives are communicated, understood and implemented for each audit.

5　Auditing

This core element gives guidance and principles addressing the issues to consider for planning and preparing for the audit, including:

- *Initiating the audit*
- *Preliminary document review*
- *Preparing the audit*
- *Audit plan, scope, and team assignments*
- *Working documents*◆

The intent, interpretation, concepts, methodologies, and techniques for EMS audit planning and preparation are found in Chapter 7.

5.3   Executing the Audit

5.3.1   Opening Meeting◆

This element is covered in Chapter 9, Pre-audit Conference.

5.3.2   Collecting Evidence◆

This element is covered in Chapter 10, The EMS Audit.

5.3.3   Audit Findings◆

This element is covered in Chapters 10 through 13, respectively.

5.3.4   Closing Meeting◆

This element is covered in Chapter 11, Post-audit Conference

5.4   Audit Reports and Records

This section addresses specific guidance on the following:

* *Audit report preparation*
* *Report content*
* *Report distribution*
* *Document retention*◆

These elements are covered in Chapter 12, EMS Audit Reporting.

6   Audit Completion

The audit is completed once all activities in the audit plan have been concluded.◆

(The author surely hopes that the audit plan includes provisions for FOLLOW-UP and CLOSE-OUT of the audit findings to assure corrective and preventive actions have been effectively implemented. ISO 14011 does not address follow-up and close-out!)

"Audit completion" is covered in Chapter 13, Follow-up and Close-out.

Summary

With respect to redundancy, the author has referenced various chapters of this book that go into complete detail (and then some) regarding the guidance provided in ISO 14011. An EMS auditor should remember that an organization may develop a more comprehensive environmental audit program based on these guidance requirements. Consideration should be given to this, especially when you take into account the limited requirements set forth in ISO 14001, Section 4.5.4, EMS audit. The remaining chapters of this book will provide suggested guidance on the overall continuous improvement effort put forth by EMS audit programme management to assure reliable, consistent, cost-effective, and comprehensive audits to effectively represent the issues regarding improved environmental performance and conformance to ISO 14001.

ISO 14012 Guidelines for Environmental Auditing—Qualification Criteria for Environmental Auditors

This guidance standard provides for the qualification criteria for environmental auditors and lead auditors. The scope extends itself beyond the competencies required for the external auditor to include internal auditors. It is interesting to note at the time of publication of this book, there is no recognized certification body or system for the certification of specifically internal environmental auditors and lead auditors. Bodies such as ANSI-RAB and EARA have strongly suggested the level and need of appropriate training in this area is equal for both external and internal auditors alike.

The guidance provided in ISO 14012 covers the following core elements and sub-elements:

- *Education and work experience*
- *Auditor training*
  - *Formal training*
  - *On-the-job training*
  - *Objective evidence of education, experience, and training*
  - *Personnel attributes and skills (Chapters 1 and 10)*
  - *Lead auditor*
  - *Maintenance of competence*
  - *Due professional care*
  - *Language*◆

ISO 14012 is additionally supported by the inclusion of Annex A (informative)—Evaluating the qualifications of environmental auditors, including:

- General information;
- Evaluation process; and
- Evaluation of education, training, and personal attributes.

Annex B (informative)—Environmental Auditor Registration Body◆

This annex contains guidance on the development of a body to ensure a consistent approach to the registration of environmental auditors.

Annex B, at the time of printing this book, is critically important informative guidance to ensure a consistent approach to the registration of environmental auditors. At present, there is limited "constructive" conversation in existence between various certification bodies, such as ANSI-RAB, EARA or

IRCA (Institute of Registered Certificated Auditors—UK). And, quite frankly, there is considerable difference between the requirements and criteria set forth by these bodies for the certification of environmental auditors and lead auditors.

Chapter 15 will address a variety of these issues and concerns. Included in this chapter is continued guidance relevant to ISO 14012, selecting the appropriate body to apply to, and key information on the application, review and registration processes, as well as several identified key differences from one body to another.

Summary

The necessity for this type of standard, by intent, is to assure a level playing field for all involved, bodies and auditors alike, for development of accreditation/certification systems, which through mutual recognition, will give the environmental professional (auditor) the freedom of choice. Ultimately, the design allows for every certified environmental and/or EMS auditor to be equally recognized regarding their qualifications in fields of environmental auditing.

3 Initial Environmental Review (IER) and Gap Analysis

Introduction

"Too many U.S. companies are confused over the difference between the terms 'initial review' and 'gap analysis,'" according to Cornelius "Bud" Smith. Smith is director of Environmental Management Services for ML Strategies Inc., a management consulting firm in Danbury, CT, and chairman of the U.S. TAG SubTAG 2 on environmental auditing.

While some experts may use the terms interchangeably, Smith explains why it is vital that companies understand the differences and the implications associated with each. According to Smith, this understanding could be the difference between going to prison or not.

Smith says every company or organization seeking to implement the forthcoming ISO 14001 EMS standard will have to conduct some variation of a gap analysis that compares its current EMS with the criteria of ISO 14001. This procedure will help it identify the system improvements necessary to achieve conformity with the standard.

Every EMS, he notes, is a web of interrelated components, including such basic elements as a policy, planning and goal setting, programs (e.g., training), measurement, auditing, reporting and communications, documentation and record keeping, and management review.

These components will vary from organization to organization depending on their relevant EMS inputs. In addition to ISO 14001, those EMS inputs may include requirements of environmental laws and regulations, industry commitments (e.g., the chemical industry's Responsible Care® initiative), and other applicable criteria such as organization internal requirements.

1. **Define your inputs**. The first steps in conducting an ISO 14001 EMS Gap Analysis are to define your organization's EMS

---

inputs and to break each of them down into a comprehensive set of meaningful subcategories.

For example, you might subdivide reporting into internal and external communications, and then again into different audiences such as management, employees, shareholders, local communities, governmental agencies, etc. For each subcategory, you should develop a list of all applicable requirements of each identified EMS input.

Since the ultimate objective is the definition of "the organization's EMS" rather than a series of individual systems related to each EMS input, you should combine the systems elements into a single set of desired organization EMS requirements for each subcategory.

When the organization compares its desired EMS requirements with its current EMS requirements, the differences identified for each subcategory will be the "gaps," or the product of the gap analysis, which will have to be addressed by the organization in order to conform to ISO 14001. The amount of effort and the degree of difficulty of conducting a gap analysis will vary depending on the complexity of the organization and the quality of its current EMS.

2. **Define your status**. To conduct a gap analysis, an organization must use some means to define the framework, criteria, and requirements of its current EMS—if one exists. ISO 14001 is silent as to an appropriate methodology for performing this task, as well as for conducting an overall gap analysis. However, some ISO 14001-related documents do provide some useful guidance—for companies with and without an EMS.

    Section A.4.3.1 of Annex A of ISO 14001 suggests that an organization without an existing EMS should undertake a "review" in the following four areas:

    (a) *legislative and regulation requirements;*
    (b) *an identification of significant environmental aspects;*
    (c) *an examination of all existing environmental management practices and procedures; and*
    (d) *an evaluation of feedback from the investigation of previous accidents.*

    Annex A recommends that the review address abnormal as well as normal operations, including potential emergencies. It also identifies "*checklists, interviews, direct inspection and*

measurement, results of previous audits, and other reviews..."
as useful tools for conducting such a review.

3. **Use initial reviews as tools—Remember, they're not required**. The European Eco-Management and Audit Scheme (EMAS) defines an "environmental review" as "initial comprehensive analysis of the environmental issues, impact, and performance related to activities at a site." In Article 3(b), EMAS requires participants to conduct such an initial review, which must include an examination of the 12 environmental issues contained in its Annex I, Part C.

   In Item 1 of Article 5, EMAS also states that registrants must base their first required public environmental statement on the results of the required initial review, and, thereafter, on required environmental audits.

   While these initial review-related provisions of Annex A of ISO 14001, ISO 14004, and EMAS are excellent references for organizations wishing to define their current EMS, they are in no way binding on an organization seeking to demonstrate conformity with ISO 14001.

   Initial reviews are required neither expressly nor by implication by the ISO 14001 specification. However, if your organization is seeking to implement ISO 14001 and does not have a formal EMS or has a very immature EMS, you might choose to use some form of an initial review as part of the gap analysis to help you define your EMS inputs and your current EMS.

4. **Tread with caution—Legal marshes may lie ahead**. Should your organization choose to conduct an initial review, a word of caution is necessary. After a look at the previously cited initial review references, it is reasonably clear that you should include an examination of your organization's compliance with applicable environmental laws and regulations within the scope of a gap analysis or initial review.

   If your company does not do any periodic environmental compliance or systems auditing, a good possibility exists that an initial review including such an audit might uncover previously unknown noncompliance situations. Accordingly, you would be well advised to consult with your attorneys about the feasibility of conducting the legal compliance segment of your initial reviews under an attorney/client privilege.[1]

Inside ISO 14004 EMS—General Guidelines on Principles, Systems, and Supporting Techniques◆

The role of the EMS auditor may begin at the earliest stages of the organization's efforts to design, document, and implement their EMS. Establishing a management process to audit and review the EMS development process at its infancy is essential. Achieving and maintaining sound environmental performance requires organizational commitment to a strong oversight and assessment process to ensure dedication to continual improvement. This requires an ongoing evaluation of practices, procedures, and processes to effectively initiate, improve, or sustain an EMS.

Inside ISO 14004

The general purpose of this guidance standard is to provide assistance to organizations implementing or improving an EMS. It is consistent with the concept of sustainable development—and is compatible with diverse cultural, social, and organizational frameworks.

This Guideline outlines the elements of an EMS and provides practical advice on implementing or enhancing such a system. Such a system is essential to an organization's ability to anticipate and meet its environmental objectives and to ensure ongoing compliance with national and/or international requirements.

The introduction to this guidance document provides for key insight into the effective concepts, principles, intent, and interpretation of an EMS, including:

- *key principles for management;*
- *benefits of having an EMS;*
- *definitions;*
- *EMS principles and elements;*
- *the Rio Declaration on Environment and Development; and*
- *ICC Business Charter for Sustainable Development.◆*

For the EMS auditor to better understand what an EMS is, how it will affect the organization's internal and external operations, and the critical role we play, a comprehensive review of this guidance document is essential.

Key Principles for Management

The design of an EMS is an ongoing and interactive process. Constant monitoring and assessment of these principles should be conducted on a regular basis to determine the effectiveness of its design. The author feels very strongly about the many hats an auditor may wear during the various phases of program development. For this reason, it is imperative we go "inside" the true spirit of commitment set forth by management in developing their EMS.

How has executive management allocated sufficient resources to the development of the organization's EMS?

Each bullet point represented in the introduction/overview section of the guidance standard on key principles for managers should address the who, how, where, and when of the statement provided. This information should be collected, documented, and reviewed throughout the EMS development process.

Who has the responsibility and authority for ensuring these key principles for management are achieved?

Organizations should consider the development of an EMS steering committee, comprised of the key management, supervision, and operating personnel, which will play a major role in the various development processes of the EMS. Defining and documenting these key responsibilities and authorities will have a significant impact on the future development of procedures for the EMS. Monitoring the performance of these individuals may allow a steady stream of information to be available to management which can be used in determining the following:

- commitment to program policies and objectives;
- human, technical, and financial resources;
- determination of training needs; and
- the means and time frames to establish objectives.

How are the key principles for management being achieved?

Review the progress being made by key personnel and the tasks assigned to them during the EMS development process. This review should be conducted at established intervals. Consideration should be given to when these reviews are conducted.

For example, management may decide to meet monthly on the progress being made. These reviews may be conducted one week prior to monthly steering committee meetings. This will allow for current information to be provided to management regarding progress.

Benefits of an EMS

From this section of ISO 14004, we begin to understand more about the key role auditors and reviewers play. We are an essential tool of management in collecting data and information used in determining just what these benefits are or may be. This collected information may become the basis for what we relay to interested parties internally or externally on our commitment and dedication to the development, implementation, and maintenance of the EMS.

By establishing a management process to audit and review the EMS development process, an organization may provide confidence to its interested parties that:

- there exists a management effort to meet the provisions of its policy, objectives, and targets;
- emphasis is placed on prevention rather than corrective action;
- evidence of reasonable care and regulatory compliance can be provided; and
- the systems design incorporates the process of continual improvement.◆

EMS Principles and Elements

The EMS model (see Figure 3.1) follows the basic view of an organization which subscribes to the following principles:

Principle 1: Commitment and Policy

An organization should define its environmental policy and ensure commitment to its EMS.

Principle 2: Planning

An organization should formulate a plan to fulfill its environmental policy.

Principle 3: Implementation

For effective implementation an organization should develop the capabilities and support mechanisms necessary to achieve its environmental policy, objectives, and targets.

Principle 4: Measurement and Evaluation

An organization should measure, monitor, and evaluate its environmental performance.

Principle 5: Review and Improvement◆

An organization should review and continually improve its environmental management system, with the objective of improving its overall environmental performance.

With this in mind, the EMS is best viewed as an organizing framework that should be continually monitored and periodically reviewed to provide effective direction for an organization's environmental activities in response to changing internal and external factors. Every individual in an organization should accept responsibility for environmental improvements.

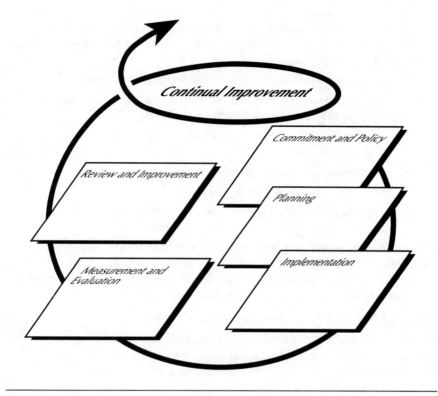

Figure 3.1 Environmental management system model.◆

Initial Environmental Review (IER)

Section 4.1.3 of the ISO 14004 guidance standard titled "Initial Environmental Review" refers to such an examination as an "environmental review" and enlarges upon its potential scope to include nine possible areas of inquiry. Let us now revisit this area from our previous reading of the ISO 14004 guidance standard.

4.1.3   Initial Environmental Review

The current position of an organization with regard to the environment can be established by means of an initial environmental review. The initial review can cover the following:

- *identification of legislative and regulatory requirements;*
- *identification of environmental aspects of its activities, products, or services so as to determine those that have or can have significant environmental impacts and liabilities;*

- *evaluation of performance compared with relevant internal criteria, external standards, regulations, codes of practice, and sets of principles and guidelines;*
- *existing environmental management practices and procedures;*
- *identification of the existing policies and procedures dealing with procurement and contracting activities;*
- *feedback from the investigation of previous incidents of non-compliance;*
- *opportunities for competitive advantage;*
- *the views of interested parties; and*
- *functions or activities of other organizational systems that can enable or impede environmental performance.*◆

In all cases, consideration should be given to the full range of operating conditions, including possible incidents and emergency situations.

The process and results of the initial environmental review should be documented and opportunities for EMS development should be identified.

Practical Help—Initial Environmental Review

An important first step is to develop the list of areas to be reviewed. This can include organization activities, specific operations, or a specific site.

Some common techniques for conducting a review include:

- *questionnaires*
- *interviews*
- *checklists*
- *direct inspection and measurement*
- *records review*
- *benchmarking*

Organizations, such as SMEs, can look to a number of outside sources including:

- *government agencies in relation to laws and permits*
- *local or regional libraries or databases*
- *other organizations for exchange of information*
- *industry associations*
- *larger customer organizations*
- *manufacturers of machinery in use*

- *business relations (e.g., with those who transport and dispose of waste)*
- *professional help*◆

In conducting the IER, we now draw your attention back to the ISO 14004 guidance document. These guidelines are intended for use as a voluntary, internal management tool and are not intended for use by EMS certification/ registration bodies as a specification standard.

Principle 1, Commitment and Policy, from ISO 14004 identifies the guiding principles for an organization's management to consider for developing their environmental policy:

- *the organization's mission, vision, core values, and beliefs;*
- *requirements of and communication with interested parties;*
- *continual improvement;*
- *prevention of pollution;*
- *guiding principles;*
- *coordination with other organizational policies (e.g., Quality, Occupational Safety & Health);*
- *specific local or regional conditions; and*
- *compliance with relevant environmental regulations, laws, and other criteria to which the organization subscribes.*

From these objectives, the guidance standard addresses some issues to be considered in environmental policy. This guidance establishes clear direction for an organization in conducting the IER.

(1) Does the organization have an environmental policy that is relevant to its activities, products, and services?

(2) Does the policy reflect the organization's values and guiding principles?

(3) Has the environmental policy been approved by top management and has someone been identified and given the authority to oversee and implement the policy?

(4) Does the policy guide the setting of environmental objectives and targets?

(5) Does the policy guide the organization towards monitoring appropriate technology and management practices?

(6) What commitments are embodied in the environmental policy (e.g., support for continual improvement, support for the

*prevention of pollution, monitor and meet or exceed legal require-
ments and consider the expectations of its interested parties)?◆*

The data and information collected from the IER on environmental policy
may now be reviewed against the "Practical Help" segment of this principle.
An organization may now clear the path for better understanding the development of a "tailor-made" environmental policy specific to their corporation's
commitment to the EMS.

Practical Help—Environmental Policy

All activities, products, or services can cause impacts on the environment.
The environmental policy should recognize this.

A detailed review of the guiding principles in Appendix A can help in
drafting an appropriate policy. The issues addressed in the policy depend on
the nature of the organization. In addition to compliance with environmental
regulations, the policy can state commitments to:

- *minimization of any significant adverse environmental impacts
  of new developments through the use of the integrated environmental management procedures and planning;*
- *development of environmental performance evaluation procedures and associated indicators;*
- *life-cycle thinking;*
- *design of products in such a way as to minimize their environmental impacts in production, use, and disposal;*
- *prevention of pollution, reduction of waste and the consumption of
  resources (materials, fuel, and energy), and commitment to recovery and recycling, as opposed to disposal, where feasible;*
- *education and training;*
- *shared environmental experience;*
- *involvement of and communication with interested parties;*
- *sustainable development; and*
- *encouragement of the use of EMS by suppliers and contractors.◆*

Summary

The IER is one of the most important preliminary steps in the development process of an EMS. Included in these processes should also be the gap
analysis of various existing Quality procedures which may be integrated with
that similar EMS requirements. For ISO 9000-registered facilities, the following elements would be ideal for the gap analysis for systems integration:

- training, awareness, and competence;
- document control;
- nonconformance and corrective and preventive action;
- EMS audit;
- management review; and
- records.

The following guidance should be considered in conducting the total initial environmental review:

- review to determine if the key principles and objectives set forth by management are being conducted as designed;
- establish controls to ensure all information relevant to Step 1 processes are reported to management on a regular basis;
- conduct IER using pre-developed criteria within the ISO 14004 guidance document *(Issues to consider 3, questionnaire).*

Note—Criteria should be modified accordingly to fit the specific needs of the organization.

- plan, schedule, conduct, report, and follow up on the IER with the same disciplines and methodologies described in Chapters 6 through 13 of this book;
- determine the extent in which change control measures are effectively monitored and implemented to the EMS development process as a result of the IER.
- conduct formal gap analysis on current and existing EMS and Quality Management System (QMS) policies and procedures; and
- establish a plan and time frame for corrective action to existing EMS and QMS policies and procedures following total IER processes.

Endnotes

1. This article was originally published in the December 1995 issue of the IESU, published by CEEM Information Services, 10521 Braddock Road, Fairfax, VA 22032, USA. Tel: +1-703-250-5900; fax: +1-703-250-4117; E-mail: <inquiry@ceem.com>; Website: <http; //www.tregistry.com/ceem/>. Reprinted with permission.

4

EMS Audit Programme Design

Introduction

An effective EMS audit is, generally speaking, only as good as the designed and implemented audit programme. The level of management commitment to this design depends on an efficient method of allocating resources, qualified personnel, and ample time to review, plan, prepare, conduct, report on results, and follow up on findings. The internal EMS audit function should be coordinated with various management organizations through the EMS representative or environmental audit manager. Because there are rarely any "full time" EMS auditors working within the internal audit programme, it is essential that this coordinated effort takes into consideration the extensive amount of time it may take to effectively conduct any EMS audit.

From ISO 14001, Annex A, on the guidance and use of the specification, Clause A.5.4 on EMS audit provides the following considerations and guidance on the audit programme and procedures:

The audit programme and procedures should cover:

(a) the activities and areas to be considered in audits;

(b) the frequency of audits;

(c) the responsibilities associated with managing and conducting audits;

(d) the communication of audit findings;

(e) auditor competence; and

(f) how audits will be conducted.

Audits may be performed by personnel from within the organization or by external persons selected by the organization. In either case, the persons conducting the audit should be in position to do so impartially and objectively.◆

55

Let us now explore each of the considerations within the Annex to better understand the concepts of developing an effective internal EMS audit program.

Activities and Areas to Be Considered in Audits

The audit programme shall consist of procedures that will define the specific guidance on how the audit function will comply with ISO 14001, clause 4.5.4-EMS Audits. The "activities" should be defined as the processes, functions, and operations within the organization. The "areas" are each activity or department in the organization to which these specific requirements are to be addressed and within which the specific requirements are to be implemented.

Document control is a specific requirement in which the total level of implementation is represented throughout the organization. Departments such as Manufacturing, Quality, Assembly, Purchasing, Stores, Shipping, Receiving, Design, and Environmental Engineering and Compliance may all contain EMS documents addressed within, and controlled by, the applicable EMS document control procedures.

ISO 14001, clause 4.5.4, requires the "audit scope" to be defined with the audit procedures. The EMS audit programme shall, by requirement, indicate how and to what extent these "activities and areas" are being considered when auditing to a given requirement.

The Frequency of Audits

Much conversation has always existed with regard to how often an organization will audit a given requirement within a management system standard. The EMS audit schedule and programme shall be based on the environmental *importance of the activity* concerned and the *results of previous audits*. Additionally, the procedures shall define the frequency based on these two key principles.

It is equally important to consider a level of flexibility when determining the frequency. And *timing* may also become a critical issue of consideration. The author recommends, on a basic approach of getting the audit program started, to conduct three EMS audits of each requirement of the ISO 14001 standard prior to the certification audit date.

The initial EMS audit should be conducted approximately 30 days after the initial effective date of the procedure being implemented within that activity. You do not want the methodologies and guidance behind controlling these disciplines to go too far "on their own" before reviewing and determining a true level of commitment to and implementation of such procedures.

Following the corrective and preventive action being addressed and implemented from the initial audit, the second audit should take place approximately 90 days following the close-out of the initial audit. This may determine the level of consistency in which the designed procedures are truly becoming an everyday part of the activities-controlled programme for conformance to specific requirements.

The third audit, and final preparatory review prior to certification by a third-party registrar, should be conducted approximately 30–60 days before the certification audit. This will allow for one last comprehensive review of the organization's ability to comply with ISO 14001 and enable the facility to better prepare for and understand the types of questions and audit process that may be covered by the registrar.

After certification, it is recommended that the organization audit each requirement of the ISO 14001 standard a minimum of twice a year. Consideration should be given to scheduling these audits prior to any annual or biannual surveillance audits conducted by your registrar.

Note—Chapters 7 and 8 on EMS audit planning and scheduling the audit, respectively, will address additional comprehensive detail in this area.

Responsibilities Associated with Managing and Conducting Audits

ISO 14001—4.5.4, EMS Audit

The organization shall establish and maintain (a) programme(s) and procedures for periodic environmental management system audits to be carried out, in order to:

(a) determine whether or not the environmental management system

(1) conforms to planned arrangements for environmental management including the requirements of this standard; and

(2) has been properly implemented and maintained.◆

There are four primary levels of responsibility and authority that should be considered in documenting and defining the key roles of personnel involved in the EMS audit programme:

1. **EMS audit programme management**

   Define and document within the EMS audit programme and procedures the key roles in providing oversight and direction to effectively implementing the requirements of the audit programme. This may, in fact, be the designated EMS representative.

2. **Top management**

   How and to what extent is top management addressed within the EMS audit programme? This may be addressed by referencing the specific policies and procedure set forth by the organization for conducting management reviews. Additionally, this referencing may allow for clear guidance on how information on the results of audits is provided to management.

3. **Management of the audited organization**

   This is addressed specifically in the body of the EMS audit reports to be distributed to the responsible management entity which will review and carry out specific corrective and preventive actions necessary based on the findings within the EMS audit report.

4. **EMS audit personnel**

   The EMS audit programme and procedures may contain several line item statements beginning with, "EMS audit personnel shall…" Review, planning, preparations, carrying out the audit, reporting, communication to the audited organization, and follow-up and close-out are certainly among the obvious elements within the audit programme that will be the responsibility of the EMS auditor. Coordination with EMS audit management of these efforts is essential and should so be defined.

Communications of Audit Findings

The requirements for EMS audits under ISO 14001, clause 4.5.4, state:

Provide information on the results of audits to management, and end the clause with…*reporting results.*

"Communication" and "reporting" should be viewed differently, typically in the sense of "verbally" versus "written."

Verbal communication of the results of EMS audits occurs at various stages throughout the audit process. Auditors often find their escorts or management of the audited organization carefully taking notes throughout the audit. Strengths and weaknesses discovered are communicated to the auditees. Daily meetings between the EMS auditors and the audited organization may also briefly disclose a summary of this information.

Post-audit activities will also communicate the various findings discovered during the audit. Immediately prior to submission of a listing of these findings, the auditor or audit team members will disclose the various findings in the post-audit conference.

The audit programme shall be based on the results of previous audits. Through this effort, a level of communication is established between the EMS auditors and EMS audit management. Because the "environmental importance of the activity concerned" is a required element, this communication is necessary. Prior to development of any audit plan or protocol, results from previous audits must be considered and discussed. Key areas and activities may require a closer look based on decisions made during this communication process.

There are additional clauses to be considered during the processes of communicating audit findings. ISO 14001, clause 4.4.3, requires *internal communication between the various levels and functions of the organization.* One may consider referencing the communication process that takes place throughout the EMS audit process in the required procedures for communication.

ISO 14001, clause 4.6-Management review, requires *addressing the possible need for changes to policies, objectives, and other elements of the EMS, in the light of EMS audit results.*

How are these results communicated to management during the review process?

While this book is not intended to be specifically on EMS program development, one must consider another element within ISO 14001. Clause 4.4.4-EMS documentation, states, *"Describe the core elements of the management system and their interaction."*

The communication of audit results is a vital element of the core continuous improvement processes within the EMS. This interaction and communication should be an integral part of specific training to EMS auditors, as well as management.

Chapter 12, EMS Audit Reporting, shall cover in detail the concepts and techniques of providing clear and concise audit reporting throughout various stages of the audit.

Auditor Competence

As you have already read up to this point, there are several key issues concerning how capable, competent, and/or technically qualified an individual is to perform effective EMS audits. Again, an extensive review of the audit processes may always be improved upon and must stay under a steady surveillance by EMS audit management. Remember, not everyone is cut out for this, no matter how much management wants them to be.

How Audits Will Be Conducted

The final line item requirements within ISO 14001, clause 4.5.4, state, *"In order to be comprehensive, the audit procedures shall cover the audit scope, frequency and methodologies, as well as the responsibilities and requirements for conducting audits..."*

The following chapters of this book shall provide the guidance and key processes to consider in order to determine how audits will be conducted. Whether your activity is specifically involved in internal, external, or third-party audits, there is certainly plenty of food for thought that can be used for making the most out of any EMS audit in which you may become involved.

5

ISO 14001
Interpretation for Audit

ISO Through Use

ISO 14001 Environmental Management Systems—Specification with guidance for use is the standard to which an organization's EMS will be audited. Whether or not an organization chooses self-declaration or certification processes, internal, second and/or third-party auditors will verify conformance to these requirements. ISO 14001 is the only specification in the current ISO 14000 series of standards containing requirements that may be objectively audited. The standard, through its core elements, establishes the "minimum" requirements to which an organization must adhere. Through effective implementation, the EMS can act as the "eyes and ears" of top management, providing information, verification, and validation of an organization's commitment to its environmental policy, objectives, and targets.

The structure of ISO 14001 contains the following sections, core elements, and subclauses:

INTRODUCTION

1   SCOPE

2   NORMATIVE REFERENCES

3   DEFINITIONS

4   ENVIRONMENTAL MANAGEMENT SYSTEM
     REQUIREMENTS

    4.1   General requirements

    4.2   Environmental policy

    4.3   Planning

        4.3.1 Environmental aspects

        4.3.2 Legal and other requirements

The core elements and subclauses of ISO 14001 will be broken down to address each line item requirement of the standard. The structure of the interpretation for audit shall contain the following, where applicable:

- Interpretation
- Annex A—Guidance on the Use of the Specification
- Common System Nonconformance
- Links between ISO 9001 and ISO 14001 (where applicable)
- EMS Auditor Checklist/Questionnaire

Interpretation

An effective EMS audit programme has, at its primary responsibility, the ability to determine the extent in which the requirements of the system, the environmental policy, and supporting procedures are implemented. An organization's programme may be distinctly different from that of another facility but have the same interpretation of the minimum requirements of this standard for both. To this end, the author has developed a comprehensive perspective into understanding how the system operates by addressing each line item requirement with the following information, where applicable:

- Key responsibilities and authorities
- Procedure and content
- Objective evidence, data, and environmental records
- Sampling attributes
- What to look for
- Common auditor mistakes

These key characteristics of the EMS audit will provide the auditor with a clear picture of a consistent and comprehensive approach to verifying system conformance.

Annex A—Guidance on the use of the specification

This annex shall provide assistance for program developers and auditors with the general interpretation and clarification of each core element and subclause. The intent of the annex is not to provide additional content to the specification. It shall essentially mirror the main body of the standard. In order to avoid any misinterpretation, the annex shall be referenced within the initial overview of the core elements and subclauses to provide direction to the user on justifying the requirement's proper definition as it may apply to any given organizational EMS.

Common System Nonconformances (CSN)

There has been much statistical evidence obtained over the years regarding management systems and the degree to which they are effectively implemented. Through this evidence, a variety of common system nonconformances shall be addressed regarding each core element and subclause. It is critical for top management and the EMS auditor, however, to remember the most important approach to effective auditing: TO LOOK FOR EVIDENCE OF CONFORMANCE. The author has chosen to address these CSNs to provide the user with an understanding of how an organization may focus its attention on typical areas in question. There is, and always will be, many an auditor who will *look for* these nonconformances because of how often they surface. The author's intention through this disclosure is simply to clarify areas that typically get pounded on during audits.

Links Between ISO 9001 and ISO 14001 Standards

ISO 9001, Quality Systems—Model for quality assurance in design, development, production, installation and servicing figures to be a major factor for organizations who will consider the integration of their existing management systems. The following core elements and subclauses of ISO 9001 and ISO 14001 contain similar requirements, thus opening the door for potential integration:

ISO 9001	ISO 14001
Training	Training, Awareness and Competence
Document and Data Control	Document Control
Corrective and Preventive Action	Nonconformance, Corrective and Preventive Action
Control of Quality Records	Records
Internal Quality Audits	EMS Audit
Management Review	Management Review

While there are distinct advantages for systems integration, there are also certain disadvantages. Defining and coordinating the efforts of different responsible individuals internally who fall under the same procedure can be difficult. Allocating "too much" responsibility to an individual or organization for managing and controlling heaps of paperwork may also lead to excessive headaches, all of which will be discussed throughout the interpretation for audit of the ISO 14001 standard.

EMS Auditor Checklist/Questionnaire

One of the most effective systems for determining what questions to ask, to whom, and at what point during the audit requires a fundamental knowledge of five basic elements within the EMS:

(1)   The requirement;

(2)   Responsibility and authorities;

(3)   Physical location the requirement(s) is/are in place;

(4)   Programmes and procedures of the EMS; and

(5)   Records (objective evidence).

The author has always recommended the technique of taking each line item statement of the requirement program or procedure and turning it around into a question.

Example: *Roles, responsibilities and authorities shall be defined, documented, and communicated....*

Auditor Questions:

How have the key *roles* within the EMS been

(a)   defined?

(b)   documented?

(c)   communicated?

How have key *responsibilities* within the EMS been

(a)   defined?

(b)   documented?

(c)   communicated?

How have *authorities* within the EMS been

(a)   defined?

(b)   documented?

(c)   communicated?

There are nine total questions an auditor may ask relevant to this line item statement. Certainly we are differentiating between roles, responsibilities, and authorities, right? And certainly we are differentiating between defined, documented, and communicated, right? These are all very different things required by paragraph 4.4.1-Structure and responsibility of ISO 14001. This section of our interpretation of ISO 14001 for audit will define and clarify the differences, as well as discuss common reference documents that may be "defined" within.

Understanding the Universal Audit Checklist (UAC)

There are key questions the EMS auditor may ask relevant to many of the core elements and subclauses of the ISO 14001 standard. The following UAC may very well come into play throughout various requirements of ISO 14001, and the author strongly recommends its use (see Chapter 7, EMS Audit Planning, for additional guidance and understanding of how the UAC fits in throughout the EMS audit process).

The following UAC addresses the disciplines of an EMS audit for collecting and obtaining information and data to be used further along in the audit:

- Document control;
- Training;
- Communication of environmental policy;
- Responsibilities and authorities implementation; and
- Records.

Universal Audit Checklist

Document control—sample three to four documents in each area audited.

Does the procedure comply with the requirements?

Are the following document control measures applied to all relevant environmental documents?

(a) *they can be located (4.4.5.A and C)*

(b) *they are periodically reviewed, revised (4.4.5.B)*

(c) *they are **approved** for adequacy by **authorized** personnel (4.4.5.B)*

(d) *obsolete documents are promptly removed or assured against unintended use (4.4.5.D)*

(e) *documentation is legible, and dates of revision are readily identifiable (4.4.5)*

Responsibilities and authorities—sample three to four.

*Are responsibilities and authorities clearly **defined** and **documented**? (4.4.1)*

Environmental policy—sample three to four.

Is the environmental policy made available to personnel (where required by policy)? (4.2)

*How has the environmental policy been **communicated** to all personnel? (4.2)*

Training—sample three to four.

What type of training have you received to perform these operations (relevant to identified "needs" and significant impacts)? (4.4.2)

Have you received environmental awareness training? (4.4.2)

What type of training have you received relevant to the requirements of ISO 14001 and applicable programmes and procedures of the EMS? (4.4.2)

What type of emergency preparedness and response training have you received? (4.4.2)

Records—sample three to four.

Are environmental records legible, identifiable, and traceable to the activity, product, or service involved? (4.5.3)

Are records and forms used and completed as designed? (4.5.3)

Let us think about this UAC for a moment. Are these not completely relevant questions to ask when auditing most of the requirements of ISO 14001? Are these five primary disciplines not verifiable throughout many locations in the facility?

Remember the one true facet of the EMS audit, we rarely have enough time to do all we need to do in the EMS audit. By collecting this data and information throughout a variety of EMS audits facility-wide, we now have obtained a thorough sample of this information. The "sample three to four" relates to the information and data gathered during an EMS audit. The key to this process relies on conducting the final EMS audits in a given cycle of these disciplines:

- Document control;
- Responsibilities and authorities;
- Training;
- Records; and
- Environmental policy.

Let us put this all together. For example, you have audited each subclause of the ISO 14001 requirements covering ten different locations throughout the facility. The following chart depicts the data gathered relevant to conducting EMS audits in a given cycle:

Auditor Sample	Discipline	Data to Collect	Sample
2–3 each area	Document Control	Document numbers, current revision; Responsibilities & authorities for approval	20–30
3–4 each area	Training	Records of names of individuals to verify training records are available for the 4 questions we ask in the UAC for training	30–40
3–4 each area	Environmental Policy	Record names of individuals to determine extent the policy has been communicated	30–40
1–2 each area	Records	Record specific environmental records used within that requirement/area	10–20
1–2 each area	Responsibility and Authority	Record and verify established roles, responsibilities and authorities are conducted as defined	10–20

NOTE—The environmental policy would need information gathered, collected, reviewed, and analyzed from other previous EMS audits to verify the organization is actively implementing subclause 4.2, A-F.

We now have an abundance of data and information gathered to effectively conduct the EMS audits for document control, training, environmental policy, records, and structure and responsibility.

This information was collected throughout the facility. Additionally, these data and this information was collected and recorded when the EMS auditor was right there to determine the extent of controls in place relevant to these requirements specific to a given location within the organization.

Through this process, the EMS audit programme may realize a reduction of audit time of up to 20%, while effectively sampling these key disciplines throughout the facility. Responsibilities and authorities (R&A) may be commonly defined and documented within programmes and procedures specific to the requirements of ISO 14001. By asking these fundamental questions in the UAC under responsibilities and authorities, the EMS audit can determine if the *documented* R&A are actually the personnel performing the task.

If using a UAC is familiar to you, then the author applauds you. If this process is new to you, the author will make every attempt at clarifying the use of the UAC in the following interpretation of ISO 14001 for the EMS audit. When the key disciplines are addressed in ISO 14001 by requirement, i.e., document control (4.4.5), training (4.4.2), environmental policy (4.2), records (4.5.3), and structure and responsibility (4.4.1), we will reflect back to the UAC for guidance in effectively conducting EMS audits to these requirements.

In summary, the author has determined that it is vital to understand this information prior to the interpretation of ISO 14001 for the EMS audit. As you probably have guessed by now, this process is equally effective for conducting ISO 9000-related Quality System Audits.

ISO 14001 Interpretation for Audit

The EMS requirements begin in section 4 of ISO 14001. Our first core element is 4.2-Environmental Policy, preceded by the general statement in 4.1:

The organization shall establish and maintain an EMS, the requirements of which are described in the whole clause 4.◆

Establish and *maintain* are terms addressed throughout the standard.

Webster's dictionary defines *establish* as: (1) bring into being, and (2) prove.

The author prefers the second definition given, "prove," which is a big part of what we as EMS auditors must accomplish throughout the EMS process. Again from Webster's dictionary, *maintain* is defined as "keep in good order...."

Well, there it is. The organization shall "prove" and "keep in good order" an EMS. And, so shall we.

We turn now to the ISO 14001, Annex A.2—Environmental policy for guidance on use.

Annex A.2—Environmental Policy for Guidance on Use

The environmental policy is the driver for implementing and improving the organization's environmental management system so that it can maintain and potentially improve its environmental performance. The policy should, therefore, reflect the commitment of top management to compliance with applicable laws and continual improvement. The policy forms the basis upon which the organization sets its objectives and targets. The policy should be sufficiently clear to be capable of being understood by internal and external interested parties and should be periodically reviewed and revised to reflect changing conditions and information. Its area of application should be clearly identifiable.

The organization's top management should define and document its environmental policy within the context of the environmental policy of any broader corporate body of which it is a part and with the endorsement of that body, if there is one.

NOTE—Top management may consist of an individual, or group of individuals, with executive responsibility for the organization.◆

ISO 14001 4.2—Environmental Policy

Top management shall define the organization's environmental policy and ensure that it

(a)   is appropriate to the nature, scale, and environmental impacts of its activities, products or services;

(b)   includes a commitment to continual improvement and prevention of pollution;

(c)   includes a commitment to comply with relevant environmental legislation and regulations, and with other requirements to which the organization subscribes;

(d)   provides the framework for setting and reviewing environmental objectives and targets;

(e)   is documented, implemented, and maintained and communicated to all employees; and

(f)   is available to the public.◆

Who is *top management*? An individual or group of individuals with executive responsibility for the organization. If the environmental policy is established for the corporate level, it may be the CEO and/or Director of Environment, Safety, and Health (ES&H). From the organization's facility, possibly the President or the designated management representative. Similar to ISO 9000 requirements for the quality policy, this environmental policy may be signed by "top management" and posted in conspicuous areas throughout the facility where the EMS is in effect. Under these conditions, the EMS auditor should verify its existence facility wide.

The most important EMS audit criteria may lie within paragraph 4.2.e, and that is that the environmental policy is documented, implemented, and maintained and communicated to *all* employees. *ALL*. Maybe the biggest "little word" in the English language. Remember the UAC from this chapter?

The EMS audit question of "How has the environmental policy been communicated to all personnel?" through the UAC process may allow the auditor to effectively "sample" how this task has been achieved. In addition, to what extent is this "communication" verifiable? If personnel indicate that they have never heard of it, is this stand-alone objective evidence that it has not been communicated? An organization should consider, while performing environmental policy and/or awareness training, documenting this communication. This can serve as objective evidence that the communication has been made to all employees.

Let us now take a look at an example of a "mock" Environmental Policy created for the purpose of instruction.

Do we notice any relationship between the requirements of ISO 14001, paragraph 4.2 and this policy? The environmental policy virtually restates all of the elements of the requirement within itself. It is tailored specifically to the organization's commitment not only to the policy, but to the EMS procedures that may assist in effectively implementing the policy.

There are additional considerations for the EMS auditor to think about to determine if the policy is effectively implemented. Take, for example, the commitment to prevention of pollution. We will observe later in this chapter the techniques for the EMS audit relevant to paragraph 4.3.3-Objectives and targets. If the objectives and targets have been verified as not being consistent with a clear commitment to pollution prevention, then we have a nonconformance against the environmental policy. Remember, the policy is the driver for implementing and improving the EMS. Again, there are various points during the EMS audit that may allow us to determine how effectively the policy is implemented.

**Quality Systems Development
(Q.S.D.)
Environmental Policy**

Q.S.D. has dedicated and committed our Environmental Management System to continually improve our organization's efforts towards the following:

- Preventing pollution;

- Complying with relevant environmental legislation and regulations;

- Complying with customer-based environmental requirements and policies; and

- Providing the framework for setting and reviewing environmental objectives and targets.

Q.S.D.'s commitment to our environmental policy has been developed specific to the organization's nature, scale, and environmental impacts of Q.S.D. activities, products, and services. This commitment is set forth in the framework of our Environmental Management System program and procedures currently implemented throughout our facilities' operations.

This policy has been communicated to all employees at Q.S.D. and is posted in conspicuous locations throughout the facility.

Our environmental policy shall be made available to the public and all interested parties upon request.

Gregory P. Johnson   *Hailey Johnson*
Gregory P. Johnson   Hailey Johnson
President/CEO-Q.S.D.   Director-ES&H-Q.S.D.

Effective date March 18, 1996. Rev.N/C Doc. No. EPS-01

The bottom of the policy contains various document control measures. Remember, the policy is a document requiring control, is it not? Which would come under the requirements of paragraph 4.4.5, right?

The environmental policy also indicates, "…[the policy] is posted in conspicuous locations throughout the facility." Because of this, we must verify that this is done as indicated. This may be done by the EMS auditor by using the UAC on various audits throughout the facility. In addition, the results of other related ISO 14001 EMS audits would support and verify whether or not the policy is effectively implemented.

The EMS auditor should also keep in mind that organizations may have various ways to assure the policy is made available to the public. Inquire into how it is made available, and review and determine if the prescribed method has, in fact, been made available to the public as designed.

Common System Nonconformances (CSN)

The following CSNs will illustrate areas of concern with regard to the effective implementation of the environmental policy. Remember, the EMS auditor is always *looking for conformance*. Special attention may be given to these areas to assure stronger visibility into the level of conformance throughout the organization's EMS:

- The environmental policy has *not* been communicated to all employees;
- The environmental policy has *not* been posted in conspicuous locations throughout the organization's facility (where applicable); and
- The environmental policy does *not* contain the appropriate document control measures, i.e., effective dates, dates of revision, and/or evidence of the authorized personnel responsible for the review, revision, and/or approval of this *document*.

The last CSN is certainly some food for thought. The environmental policy is a *document*, is it not? Maybe it is the most important document relevant to the organization's EMS. Would it not also fall under the requirements of ISO 14001, subclause 4.4.5-Document control?

The author will exit this interpretation of audit of ISO 14001, subclause 4.2-Environmental policy with this thought and/or advice.

EMS Auditor Checklist/Questionnaire

The environmental policy is the driver for implementing and improving the EMS. To this end, the environmental policy greatly differs from that of an ISO 9001 quality policy. Many other system subclauses would need to be reviewed and assessed prior to determining the effective implementation of the environmental policy. The following checklist depicts the key auditor questions for carrying out an effective determination of the adherence to the environmental policy.

- Has top management defined the organization's environmental policy?

- Is the policy:
  A. appropriate to the nature, scale, and environmental impacts of its activities, products, or services? (4.3.1)

  B. committed to continual improvement and prevention of pollution? (4.3.3 and 4.3.4)

C. committed to comply with relevant environmental legislation and regulations and other organizational requirements? (4.3.2 and 4.5.1)

D. providing a framework for setting and reviewing environmental objectives and targets? (4.3.3 and 4.3.4)

E. documented, implemented, maintained and communicated to all employees? (UAC)

F. available to the public?

The subclauses identified after each question illustrate key areas in which subsequent assessments will provide data and information critical in determining whether or not the environmental policy is effectively implemented. Our final question relates to the process for which the EMS auditor may determine the following:

Have results from previous EMS audits been reviewed to determine the effective implementation of the environmental policy?

In summary, the UAC will play a key role in gathering data and information relevant to the EMS audit to the environmental policy. Consideration should be given during the planning of each applicable EMS audit which will gather data and information relevant to the determination of the extent to which the environmental policy is effectively implemented.

ISO 14001—4.3 Planning

The critical core element within the EMS is the heart of the requirements essential to the organization's efforts in developing, documenting, implementing, and maintaining continuous improvement. From ISO 14001, Annex A, let us review A.4.3.1—Environmental aspects.

Annex A.3.1—Environmental Aspects

This section is intended to provide a process for an organization to identify significant environmental aspects that should be addressed as a priority by the organization's environmental management system. This process should take into account the cost and time of undertaking the analysis and the availability of reliable data. Information already developed for regulatory or other purposes may be used in this process. Organizations may also take into account the degree of practical control they may have over the environmental aspects being considered. Organizations determine what their environmental aspects are, taking into account the inputs and outputs associated with their current and relevant past activities, products, and services.◆

An organization with no existing environmental management system should, as a first step, establish its current position with regard to the environment by means of a review. The aim should be to consider all environmental aspects of the organization as a basis for establishing the environmental management system.

Those organizations with environmental management systems do not have to undertake such a review.◆

NOTE—See Chapter 3, Initial Environmental Review and Gap Analysis, for additional information regarding this review process.

From ISO 14004:

The review should cover four key areas:

(a) legislative and regulatory requirements;

(b) an identification of significant environmental aspects;

(c) an examination of all existing environmental management practices and procedures; and

(d) an evaluation of feedback from the investigation of previous incidents.

In all cases, consideration should be given to normal and abnormal operations within the organization and to potential emergency conditions.

A suitable approach to the review may include checklists, interviews, direct inspection and measurement, and results of previous audits or other reviews, depending on the nature of the activities.

The process to identify the significant environmental aspects associated with the activities at operating units should consider, where relevant:

(a) emissions to air;

(b) releases to water;

(c) waste management;

(d) contamination of land;

(e) impact on communities;

(f) use of raw materials and natural resources; and

(g) other local environmental issues.

This process should consider normal operating conditions, and shut down and start up conditions, as well as realistic potential significant impacts associated with reasonably foreseeable or emergency situations.

The process is intended to identify significant environmental aspects associated with activities, products, or services, and is not intended to require a detailed life-cycle assessment. Organizations do not have to evaluate each product component or raw material input. They may select categories of activities, products, or services to identify those aspects most likely to have a significant impact.

The control and influence over the environmental aspects of products varies significantly, depending on the market situation of the organization. A contractor or supplier to the organization may have comparatively little control, while the organization responsible for product design can alter the aspects significantly by changing, for example, a single input material. Whilst recognizing that organizations may have limited control over the use and disposal of their products, they should consider, where practical, proper handling and disposal mechanisms. This provision is not intended to change or increase an organization's legal obligations.◆

The following are the specific requirements within ISO 14001, sub-clause 4.3.1-Environmental aspects:

The organization shall establish and maintain (a) procedure(s) to identify the environmental aspects of its activities, products, or services that it can control and over which it can be expected to have an influence, in order to determine those which have or can have significant impacts on the environment. The organization shall ensure that the aspects related to these significant impacts are considered in setting its environmental objectives.

The organization shall keep this information up-to-date.◆

The organization shall *prove*, and *keep in good order*, a procedure...you mean a *documented* procedure, right? Hmm. Can you think of any reason why this word "documented" was left out of so many places in this specification? Considering the unbelievable events that actually surfaced regarding the ISO 9000 certification process *before* they changed the standard in 1994, the author is certainly puzzled. There were actually several facilities who were *certified* to the management system requirements of ISO 9001/2 *without* having more than a half-dozen documented procedures. For management personnel and the EMS auditor with us today, let us ponder this question:

Can an organization have an effectively implemented, understood, and consistently carried out EMS without documented procedures for defining how this may be accomplished?

There may be a few of you who are saying, "Yes." Really?

Ladies and gentlemen of the jury, I leave you with this...since the beginning of mankind the necessity for having a procedure implies a *documented procedure*. This is and always will be the *intent* regarding the term "procedure" relevant to management system standards and specifications.

Still in doubt? Then please check with your registrar for their interpretation relevant to what they will be looking for in a "procedure." For the remainder of the ISO 14001 interpretation for EMS audit, the author has put the invisible word, "documented," before each of the places in the specification that reference a procedure. For those of you allocated the responsibility for technical development of your organization's EMS procedures—heed the warning. If the organization does not develop "documented" procedures, the EMS auditor must determine that there is a "controlled and consistent process in place" to satisfy a given requirement.

"...a procedure to identify the environmental aspects of its activities, products or services." Identify. The EMS auditor should determine how the organization carries out the process for identifying via the procedure.

Environmental aspects. To better understand this term, we must fast-forward to ISO 14001, subclause 4.3.3-Objectives and targets. This requirement addresses *significant* environmental aspects. *Significant.* Is there a difference between significant environmental aspects and just plain old environmental aspects? You bet! The identified "significant aspects" are those in which the organization has chosen to establish, identify, and document their objectives and targets. The EMS auditor should determine the interaction between the identified significant aspects and the objectives and targets established by the organization.

There has been much debate regarding the role of the EMS auditor, internally or externally, in defining just what an environmental aspect is. For example, let us say emissions to air is a significant environmental aspect, in the opinion of the auditor. But the organization being audited does not address this issue as a significant aspect. There is currently no evidence of established and documented objectives and targets relevant to this significant aspect. Is this a systems nonconformance? Remember, the EMS auditor is evaluating the *organization's* EMS, not the opinion of what that auditor believes one should be. It is our objective to determine how the organization goes about identifying environmental aspects that it determines are relevant to their environmental policy and their facility's specific operations.

Let us take this point one step further. At the post-audit conference/exit interview, the EMS auditor has indicated that this is a systems nonconformance for not identifying the emissions to air as a significant environmental aspect. The audited organization then explains the following: Show me in ISO 14001, subclause 4.3.1 (environmental aspects) or 4.3.3 (objectives and targets) where it says we must identify air emissions as a significant environmental aspect. It is not there. Gotcha! The EMS auditor has been challenged. Again, we are auditing the organization's defined significant environmental aspects, and an EMS auditor does not make a self-determination of what is or is not *significant*. But remember, if the organization does not identify air emissions as an environmental aspect, the auditor will react.

"...the environmental aspects of its activities, products or services."

From ISO 14004 EMS—Guidelines on Principles, Systems, and Supporting Techniques, subclause 4.2.2-Identification of environmental aspects and associated environmental impacts, addresses the following:

Activity	**Aspect**	**Impact**
handling of hazardous materials	potential for accidental spillage	contamination of soil or waste
Activity	**Aspect**	**Impact**
a product	reformulation of the product to reduce its volume	conservation of natural resources
Activity	**Aspect**	**Impact**
vehicle maintenance	exhaust emissions	reduction of air pollution

The EMS auditor should inquire as to how the organization has taken into consideration each of the three disciplines—activity, product, or services—in identifying their environmental aspects.

Determination of Significant Impacts

From ISO 14004, subclause 4.2.2:

Step 4: Evaluate Significance of Impacts

The significance of each of the identified environmental impacts can be different for each organization. Quantification can aid judgement.

Evaluation can be facilitated by considering:

Environmental Concerns
- *the scale of the impact;*
- *the severity of the impact;*
- *probability of occurrence; and*
- *duration of impact.*

Business Concerns
- *potential regulatory and legal exposure;*
- *difficulty of changing the impact;*
- *cost of changing the impact;*

> * *effect of change on other activities and processes;*
> * *concerns of interested parties; and*
> * *effect on the public image of the organization.*◆

The EMS auditor should determine how the organization has determined what their significant impacts are relative to their identified environmental aspects that it can control and over which it can be expected to have an influence. By inquiring as to their knowledge and understanding of what their organization's significant environmental impacts are, the auditor will be lead to a better understanding of the organization's logic behind how they have identified their significant environmental aspects.

This will, in turn, lead to the EMS auditor's determination and evaluation of how the organization considered these significant impacts in setting its objectives and targets.

"...the organization shall keep this information up-to-date."

Significant changes to an organization's activities, products, or services, including related environmental legislation or modifications thereto, play a major role in when and how this information will be kept up-to-date. The document control requirements of 4.4.5 come into play regarding the assurance of how effectively this information is controlled.

With respect to keeping the procedure up-to-date, keep in mind this procedure should address controlled methods for change control. And, by the way, how do you keep a procedure up-to-date that is not documented? Oh, sorry, we went over that already, did we not?

Common System Nonconformances (CSN)

* Environmental aspects have *not* been identified of its activities, products, and services.
* Environmental aspects have *not* been identified that the organization can control or over which it can be expected to have an influence.
* Information is *not* kept up-to-date.

With respect to our second CSN, remember, we are distinguishing a difference between environmental aspects and *significant* environmental aspects. Take, for example, our discussions on emissions to air. Because of local, state, and/or federal air quality regulations and standards, a facility is expected, as required, to control and have an influence over these emissions to air. However, the organization may have *initially* chosen not to identify this as a *significant* environmental aspect. Therefore, there may not *initially* be documented environmental objectives and targets related to this environmental aspect within the EMS. Technological options; financial, operational, and business requirements; views of interested parties; and legal and other requirements will play a major role in the organization's establishment and determination of what their significant environmental aspects are.

EMS Auditor Checklist/Questionnaire

- Has the organization established and maintained a procedure to identify the environmental aspects of its
  A. activities?
  B. products?
  C. services?
- Is the procedure effectively implemented?
- Have environmental aspects been identified for its activities, products, or services
  A. that it can control?
  B. over which it can be expected to have an influence?
- How has the organization determined the significant impacts on the environmental for which it
  A. has
  B. or can have control over?
- How has the organization ensured that the aspects related to these significant impacts are considered in setting its environmental objectives?
- Are document control measures in place to keep this information up-to-date, including change control?

ISO 14001, 4.3.2—Legal and Other Requirements

The organization shall establish and maintain a procedure to identify and have access to legal and other requirements to which the organization subscribes, that are applicable to the environmental aspects of its activities, products, or services.◆

From ISO 14004, subclause 4.2.3-Legal and other requirements:

To maintain regulatory compliance, an organization should identify and understand regulatory requirements applicable to its activities, products, or services. Regulations can exist in several forms:

- *those specific to the activity (e.g., site operating permits);*
- *those specific to the organization's products or services;*
- *those specific to the organization's industry;*
- *general environmental laws; and,*
- *authorizations, licenses and permits.*

Several sources can be used to identify environmental regulations and ongoing changes, including:

- *all levels of government;*
- *industry associations or groups;*
- *commercial databases; and,*
- *professional services.*

To facilitate keeping track of legal requirements, an organization can establish and maintain a list of all laws and regulations pertaining to its activities, products, or services.◆

To successfully implement this requirement, it requires a strong document control program for managing, keeping up-to-date, and controlling a plethora of environment-related documents, data, regulations, and standards. Document control? Wait a minute. This is an overview for the EMS auditor relevant to legal and other requirements, is it not? Indeed it is. So let us ponder these next two questions:

- Are legal and other requirements documents pertinent to the organization's EMS?
- Does this mean ISO 14001, subclause 4.4.5-Document control, relates directly to these "legal and other requirements"?

Yes, *of course*. Shocked by this interpretation? Let us fast forward to the first sentence of ISO 14001, subclause 4.4.5-Document control:

The organization shall establish and maintain procedures for controlling ALL DOCUMENTS REQUIRED BY THIS STANDARD....

Quickly, throw it in reverse: "The organization shall establish and maintain a procedure to *identify* and *have access to...*"

Legal and other requirements are found in documents, are they not? Hmm.

The EMS auditor must remember, however, that he or she is auditing the organization's EMS. How has the organization defined and documented controls over legal and other requirements documentation? To effectively audit this requirement dictates careful planning and sampling.

"...a procedure to identify..." How has the organization identified these legal and other requirements? Perhaps, for example, there is a detailed listing contained within their computer databases compiling and identifying all of these requirements. How is this information controlled and kept up-to-date? To verify this, the auditor may pull a sample of a variety of legal and other requirements depicted by the following chart.

Sample	Type of Requirement	Total Sample
3–4 each	local, state, federal, permits	12–16

The EMS auditor should document the title of the requirement and its current revision. From there, inquire as to how the organization has "access to" these requirements.

Access to. Does this mean that every requirement must be readily available at this location? Physically?

What if their absence could lead to deviations from the environmental policy and the objectives and targets?

For example, if an operator was performing waste stream classification and certification in accordance with the Resource Conservation and Recovery Act (RCRA), should these requirements be readily available to the operator? To the facility in whole?

If the facility can show that these requirements are accessible to the facility's personnel, then the EMS auditor has verified what the requirement states—*access to*.

But how long must an auditor wait to determine that they have access to these requirements? If the EMS auditor has not been able to verify access to these requirements within one working day, is this an EMS nonconformance?

Documentation which is not accessible at the time of the audit raises a concern. How effectively can the organization *implement* these requirements if they do not have reasonable access to them, especially if they have a known impact on operations essential to the effective implementation of that requirement or those requirements of the EMS, policy, objectives, and targets? Under these conditions, the auditor must "call it as it is"—not available at the time of the audit. Remember, the UAC may play a major role in this EMS audit regarding the document control measures we are always looking for.

Annex A of ISO 14001 contains the following types of "other requirements."

Annex A.3.2—Legal and Other Requirements

Examples of other requirements to which the organization may subscribe:

(a) Industry codes of practice;

(b) Agreements with public authorities; and

(c) Nonregulatory guidelines.◆

Common System Nonconformances (CSN)

- **The procedure does not address how the "other requirements" to which the organization subscribes have been identified.**

- **Legal and other requirements (based on auditor sample) are not readily accessible at the time of the audit.**

EMS Auditor Checklist/Questionnaire

- Has the organization established and maintained a procedure to

   A. identify legal and other requirements to which the organization subscribes?

   B. have access to legal and other requirements to which the organization subscribes?

   C. are the organization's legal and other requirements applicable to the environmental aspects of its activities, products, or services?

ISO 14001, 4.3.3—Objectives and Targets

The organization shall establish and maintain documented environmental objectives and targets at each relevant function and level within the organization.

When establishing and reviewing its objectives, an organization shall consider the legal and other requirements, its significant environmental aspects, its technological options, and its financial, operational, and business requirements, and the views of interested parties.

The objectives and targets shall be consistent with the environmental policy, including the commitment to prevention of pollution.

Annex A.3.3—Objectives and Targets

The objectives should be specific; targets should be measurable wherever practicable, and, where appropriate, take preventative measures into account.

When considering their technological options, an organization may consider the use of best available technology where economically viable, cost effective, and judged appropriate. The reference to the financial requirements of the organization is not intended to imply organizations are obliged to use environmental cost accounting methodologies.

The requirements for setting, establishing, and documenting environmental objectives and targets is itself at the heart of the continuous improvement requirements regarding environmental performance. To better illustrate and understand the types of environmental objectives and targets committed to by an organization, ISO 14004, subclause 4.2.5, includes the following:

Objectives can include commitments to:

- *reduce waste and the depletion of resources;*
- *reduce or eliminate the release of pollutants into the environment;*
- *design products to minimize their environmental impact in production, use and disposal;*
- *control the environmental impact of raw material sourcing;*
- *minimize any significant adverse environmental impact of new developments; and,*
- *promote environmental awareness among employees and the community.*

Progress towards an objective can generally be measured using environmental performance indicators such as:

- *quantity of raw material or energy used;*
- *quantity of emissions such as CO_2;*
- *waste produced per quantity of finished product;*
- *efficiency of material and energy use;*
- *number of environmental incidents/accidents;*
- *% waste recycled;*
- *% recycled material used in packaging;*
- *number of vehicle kilometers per unit of production;*
- *specific pollutant quantities, e.g., NO_x, SO_2, CO, HC, Pb, CFCs;*
- *investment in environmental protection;*
- *number of prosecutions; and,*
- *land area set aside for wildlife habitat.*◆

---

*"...**documented** objectives and targets at each relevant function and level within the **organization**."*

The EMS auditor, prior to the audit, should carefully plan each stage of the audit to determine the effectiveness of the design of how objectives are determined and how targets are achieved. In verifying this, the objectives and targets "program" may be viewed by the auditor during three different phases of the program:

- Initial design and development
- Midstream
- After completion of achieving set targets

Initial Design and Development

The EMS auditor should review the processes in which the organization initially establishes, designs, and documents their environmental objectives. Determining responsibility and authorities for achieving this task is essential. Through a "walk through" of this design process, the EMS auditor should verify how each relevant function and level within the organization is affected. For example, an organization may elect to document an environmental objective of using an alternative type of raw material to be incorporated into a new or modified product line. The ultimate goal (target) may lead to a minimization of pollution from their production processes, while still satisfying the requirement and the needs of their customers. The following chart below illustrates relevant functions and levels within the organization that may be affected by this environmental objective:

Function	Level	Responsibility
Purchasing	Management	Review and determination of key suppliers to provide alternative raw material, including oversight, budgeting, and availability.
Purchasing	Buyers	Supplier contact, pricing. Obtaining and purchasing samples for analysis, design impact.
Quality Control	Supervisor	Oversight of receiving process, material control, and alternative inspection/test methods.
Quality Control	Inspectors	Inspect, test.

Additional functions within an organization which may be affected by this environmental objective include, but are not limited to:

- Design;
- Environmental Engineering/Management;

- Manufacturing;
- Shipping/Receiving;
- Sales/Marketing;
- Human Resources (Training).

From these examples, the EMS auditor should remember that many of the established environmental objectives may be distinctly different from each other. Many of the functions listed above may not be "affected" by the design of the documented environmental objectives. By requirement, the organization shall review its objectives. The EMS auditor verifies how this review takes place and how the organization *considers* the following:

- Legal and other requirements;
- Significant environmental aspects;
- Technological options;
- Financial, operational, and business requirements; and
- Views of interested parties.

Each of these considerations may be reviewed by different "relevant functions" within the organization. In setting and establishing these objectives, EMS auditors may find themselves investigating, questioning, and reviewing each relevant function and how these considerations are reviewed by the appropriate function.

Midstream

Still using our raw material alternative example above, the EMS auditor may be in a position to evaluate the effectiveness of how the documented environmental objectives are being implemented. Again, each relevant function involved in the design of the objective is investigated to assure that they are carrying out their end of the process. A critical determination for the EMS auditor to consider now becomes change control.

To what extent has the original design of the documented objective been subject to modification and change, how does this impact each relevant function, and how does the organization control these changes? The UAC covering the elements of document control must be taken into consideration.

After Completion of Achieving Targets

The EMS auditor may also be in the position to conduct a complete assessment from beginning to end of how an environmental objective was designed, documented, and implemented, including the achievement of the set target. This type of audit may be the most valuable information for management to review and consider in determining strengths and weaknesses of the overall environmental management program established for achieving its objectives and targets. As the EMS further develops and integrates itself within the organization, the EMS audit program may consider conducting an integrated EMS audit covering both the objectives and targets *and* environmental management program(s) requirements of ISO 14001.

Whether conducting an EMS audit during initial design and development, midstream, or after completion of achieving targets, the environmental objectives, by design, *must be consistent with the environmental policy, including the commitment to prevention of pollution.* The EMS auditor will review the policy against set objectives and targets to determine that the dedication and commitment to implementing their policy is consistent with the objectives and targets being audited. In addition, the EMS auditor must verify that an objective and target has been established and documented concerning itself with the prevention of pollution.

Common System Nonconformances (CSN)

The EMS audit function must take into consideration the following: has a *reasonable* approach been taken into the development of environmental objectives and targets? Organizations may initially set goals for themselves that are not initially obtainable by design. Establishing and setting too many environmental objectives and targets initially could lead to insufficient allocation of resources, including human resources, specialized skills, and training, financial, and technological resources. Additional CSNs include:

- Inadequate change control processes built in to the design of the objectives and targets.

- The organization has *not* effectively considered all related requirements, options, and/or significant environmental aspects in establishing and carrying out their objectives.

- Relevant functions and levels within the organization are *not* adequately established and defined, including communication of responsibilities and authorities.

EMS Auditor Checklist/Questionnaire

- Has the organization established and maintained documented environmental objectives and targets?
- How is each relevant function and level within the organization documented within the environmental objectives and targets?
- Are these objectives effectively being implemented by design? (Sample 3 or 4 relevant, defined, and documented functions)
- How does the organization review its objectives?
- How does the organization consider the following when establishing and reviewing its objectives:
    A. Legal and other requirements?
    B. Significant environmental aspects?
    C. Technological options?
    D. Financial requirements?
    E. Operational requirements?
    F. Business requirements?
    G. Views of interested parties?
- Are the objectives and targets consistent with the environmental policy?
- Is the organization's commitment to prevention of pollution established within their objectives and targets?

ISO 14001, 4.3.4—Environmental Management Programme(s)

The organization shall establish and maintain (a) programme(s) for achieving its objectives and targets. It shall include:

(a) designation of responsibility for achieving objectives and targets at each relevant function and level of the organization; and

(b) the means and time frame by which they are to be achieved.

If a project relates to new developments and new or modified activities, products, or services, programme(s) shall be amended where relevant to ensure that environmental management applies to such projects.◆

Annex A.3.4—Environmental Management Programme(s)

The creation and use of one or more programme(s) is a key element to the successful implementation of an environmental management system. The programme should describe how the organization's objectives and targets will be achieved, including time frames and personnel responsible for implementing the organization's environmental policy. This programme may be subdivided to address specific elements of the organization's operations. The programme should include an environmental review for new activities.

The programme may include, where appropriate and practical, consideration of planning, design, production, marketing, and disposal stages. This may be undertaken for both current and new activities, products, or services. For products, this may address design, materials, production, processes, use, and ultimate disposal. For installations or significant modifications of processes, this may address planning, design, construction, commissioning operation and, at the appropriate time determined by the organization, decomissioning.◆

Planning is the key element in addressing the organization's environmental objectives. The environmental management programmes should address the organization's strategic plan for scheduling, allocating resources, and determining key responsibilities and authorities for achieving their environmental objectives and targets. These specific actions assist the organization in improving its environmental performance within individual processes, projects, services, and products.

The first objective for the EMS auditor is to verify the established responsibilities for achieving objectives and targets. After verifying the establishment of such responsibilities, the EMS auditor will determine the competence and effectiveness of the programme objectives at a sample of each relevant function and level within the organization. Specific audit protocol and checklists should be developed against the programme to determine the extent to which management, supervision, and line personnel understand these objectives and the effectiveness of the programme implementation.

The *means* by which they are to be achieved will set forth specific tasks that key personnel will perform in order to facilitate the programme objectives. Again, audit protocol and checklists should be developed against the environmental management programmes to verify key personnel are effectively implementing their part of the programme's design. *Time frames* to complete specific tasks for each relevant function is a critical part of the EMS audit function. Have key personnel been allocated enough time to perform these tasks? Are they aware of such time frames for completion? To this end, the EMS auditor must determine the processes in place to effectively *communicate* the status of programme objectives, including the processes for *change*. It only takes one of these relevant functions getting behind schedule to have a definite impact on achieving their objectives and targets and the priorities schedule of another function.

Environmental management programmes may have a distinct impact on new developments and new or modified activities, products, or services. This information should be determined during the planning and preparation stages of the EMS audit. To be effective, the EMS auditor must determine the impact that the programme has on new developments and how this has been addressed, communicated, and understood by key personnel within each relevant function within the organization. Evidence within the programme of effective measures of change control should be verified to allow for the integration of the environmental management programme plan within the specific new or modified activities, products, or services.

As discussed previously, the EMS auditor should also determine the extent to which the organization has established "reasonable" objectives and targets. There may be several key personnel who are working within more than one program. Prioritization becomes a verifiable element of the program's design. Can a higher priority programme have an adverse impact on other environmental management programmes facilitated by the same activity? Again, change control becomes a critical element of the programme's design. The UAC addresses a variety of document and change control verification questions and methods essential to performing an effective EMS audit of the environmental management programme requirements.

There is an important factor for the EMS auditor to consider when auditing this requirement. "Things" happen (the author has slightly changed the wording to this popular bumper sticker). Does an EMS nonconformance exist if the organization does not meet or exceeds the time frame for achieving its target? If the following conditions exist regarding modification of set time frames, then an EMS nonconformance does not exist:

- Documented measures of effective change control.
- Communication measures on the impact this modification will have on all relevant functions and levels within the organization for the specific environmental management programme.
- Verification of the restructuring of provided resources (i.e., human, specialized skills, technological, and financial) essential to the implementation and control of the environmental management programme.

Much to consider. With respect to redundancy, the author wishes to draw your attention to the above regarding our listing of common system nonconformances. Careful attention should also be made regarding the *ISO 14001 Interpretation for Audit* of our next subclause, 4.4.1-Structure and responsibility. There are many key EMS audit concepts within this interpretation that will play a major role in this audit, as well as other related ISO 14001 EMS audits.

In summary, the EMS auditor should remember our sampling processes. It is safe to say you may not have time to audit all of the environmental management programmes in place within the organization. In sampling the system, the EMS audit function should consider auditing two established programmes in which one has a direct impact on new developments or new or modified activities, products, or services. The next time the EMS audit is to be conducted on environmental management programmes, you may consider auditing two different programmes, then the initial audit, and so on.

Common System Nonconformances (CSN)

- **The means by which environmental objectives and targets are to be achieved are not effectively documented and communicated to key personnel at each relevant function within the organization.**

- **Environmental management programmes have not considered the new developments and/or modifications to activities, products, or services within the programme design.**

EMS Auditor Checklist/Questionnaire

NOTE—This author wishes to inform our readers at this time that it could take as long as two to three working days to **effectively** audit two different programmes, including preparation, planning, protocol/checklist development, and audit. Food for thought!

- Has the organization established and maintained (a) programme(s) for achieving its objectives and targets? (Sample two programmes)

- Are responsibilities for achieving objectives and targets being designated for each relevant function and level of the organization?

- How have the means for achieving objectives and targets been established?

- Are these means effectively implemented? (Sample three or four functions)

- Have the time frames been established for achieving objectives and targets? (Sample three or four functions)
- How has the programme(s) been amended where relevant to ensure that environmental management applies to
    A. new developments, activities, products, or services?
    B. modified activities, products, or services?

ISO 14001, 4.4—Implementation and Operation

4.4.1—Structure and Responsibility

Roles, responsibility and authorities shall be defined, documented, and communicated in order to facilitate effective environmental management.

Management shall provide resources essential to the implementation and control of the environmental management system. Resources include human resources and specialized skills, technology, and financial resources.

The organization's top management shall appoint (a) specific management representative(s) who, irrespective of other responsibilities, shall have defined roles, responsibilities, and authority for

(a)   ensuring that environmental management system requirements are established, implemented, and maintained in accordance with this international standard; and

(b)   reporting on the performance of the environmental management system to top management for review and as a basis for improvement of the environmental management system.◆

Annex A.4.1—Structure and Responsibility

The successful implementation of an environmental management system calls for the commitment of all employees of the organization. Environmental responsibilities, therefore, should not be seen as confined to the environmental function, but also include other areas of an organization, such as operational management or staff functions other than environmental.

This commitment should begin at the highest levels of management. Accordingly, top management should establish the organization's environmental policy and ensure that the environmental management system is implemented. As part of this commitment, the top management shall designate specific management representative(s) with defined responsibility and authority for implementing the environmental management system. In large or complex organizations, there may be more than one designated representative. In small or medium enterprises, these responsibilities may be undertaken by one individual. Top management should also ensure that an appropriate level of resources are provided to ensure that the environmental management system is implemented and maintained. It is also important that the key environmental management system responsibilities are well defined and communicated to the relevant personnel.◆

Certainly one of the more difficult areas for the EMS audit function to effectively verify is whether or not sufficient resources have been allocated facility-wide to implement the EMS. During our pre-audit preparations and planning, the EMS auditor should determine how the organization has determined just what are sufficient resources relevant to each level and function within the organization responsible for implementation of the EMS. By sampling the various elements within the EMS, the EMS auditor should break down the three primary levels of roles, responsibilities, and authorities for a given function.

These include:

- Management
- Supervisors
- Key operational personnel

After determining the resources (time frames, personnel, equipment), the EMS auditor should objectively verify, question, and investigate whether or not these predetermined resources have, in fact, been allocated to key personnel in order to give them sufficient time to complete tasks relevant to the effective implementation of the EMS. Is there consistent, verifiable objective evidence between what management has allocated and the resources *actually* available to and for personnel responsible for implementation? To effectively determine these factors, let us further examine and define who the key players are relevant to the ISO 14001 requirements.

ROLES—Tasks individuals or groups are required to perform to effectively implement the EMS.

RESPONSIBILITIES—Individuals or groups held accountable for performing these tasks as designed.

AUTHORITIES—Individuals or groups having designated "official power" to see to it that the overall system requirement(s) are effectively carried out.

These are three very different things, are they not? Because of this difference, the EMS auditor must verify *first* that these roles, responsibilities, and authorities are defined. To these definitions, they must also, in turn, be *documented*. Through careful review during the planning and preparation of a given EMS audit, protocol should be developed for the EMS auditor to verify these defined and documented roles, responsibilities, and authorities are carried out as designed.

Remember the UAC? This is a common task that the EMS auditor must perform when auditing any requirement of the ISO 14001 specification. But for specifically auditing this requirement, a larger sample of the entire EMS may be reviewed to ensure that these key individuals have clearly defined and documented roles, responsibilities, and authorities.

The next task for the EMS auditor is to determine how these defined and documented roles, responsibilities, and authorities have been communicated to key personnel.

Again, back to the UAC. Understanding and verifying how the given requirements have been communicated to key personnel is relevant to all EMS audits for ISO 14001. But specific to auditing this requirement, the EMS auditor should question the "authority" over a given requirement/ function to determine what steps were taken to communicate this information to key personnel.

Throughout this overview, we have engaged in much discussion regarding the UAC and the connection with this requirement and other EMS audits. Take into consideration four of the most common responses an EMS auditor may hear when the potential of an EMS nonconformance exists:

- "We don't have enough **time** to complete this task."
- "Our activity has not been allocated sufficient **funding** to carry out these tasks."
- "We don't have the proper **equipment** to effectively do this here."
- "We don't have enough **people** to do all of this with the heavy workload here."

Time. Money. Equipment. Personnel.... Hmm.

From these examples, think about this:

During an EMS audit of ISO 14001, 4.3.4-Environmental management programme(s), the following EMS nonconformances were detected at the time of the audit:

- Program requirements have not been communicated and carried out by the following functions identified in the programme:

    A. Purchasing

    B. Shipping/Receiving

    C. Raw Material Storage

- Analytical testing of raw material samples have not been performed at the time of the audit in accordance with programme requirements and within specific designated time frames for completion.

- Unauthorized changes were made to programme requirements by the following functions, having a potential significant impact on establishing and achieving the set target for this programme:

    A. Purchasing

    B. Laboratory (testing)

    C. Marketing/Sales

A significant number of EMS nonconformances have been detected and discovered during this audit.

We have another EMS nonconformance, do we not?

Can the EMS auditor now draw a conclusion that roles, responsibilities, and authorities have not been effectively communicated to relevant functions?

Have essential resources been provided to facilitate effective environmental management?

Based on objective evidence discovered during this example EMS audit, we now have an EMS nonconformance against ISO 14001, 4.4.1-Structure and responsibility.

This is a very important factor for the EMS auditor and audit organization to consider. If significant EMS nonconformances are found relevant to any ISO 14001 EMS audit, then there are not *essential* resources being provided to effectively facilitate environmental management.

Let us look at it another way.... An EMS audit to ISO 14001, 4.3.4-Environmental management programme(s), has found there are no EMS nonconformances. Can we now say that resources essential to the implementation and control of the EMS for this requirement are effective? You bet!

There is a great deal of required interaction between different requirements of the ISO 14001 specification. So much, in fact, that a specific requirement has been developed to ensure this interaction (4.4.4-EMS documentation).

In light of this information, we now begin, hopefully, to better understand the use of UAC in determining appropriate levels of interaction existing between different functions and requirements of ISO 14001 and the designed EMS.

...The organization's top management shall appoint (a) specific management representative(s)....

Does this representative have to be a member of management within the organization? If you say the answer is "No," especially third-party auditors, take into consideration the following interpretation.

By requirement, this management representative has been designated the authority *for ensuring*. The authority. By definition (from Webster's Dictionary) the "official power." The official power for ensuring the EMS requirements are established, implemented, and maintained in accordance with this standard. Let us think about this....

A nonmanagement representative has the official power for ensuring management is effectively carrying out its responsibilities and authorities of the EMS and is reporting back to management to ensure continuous improvement? Get real!

Still not convinced? Then why did the ISO 9001 Quality System Standard change this requirement in 1994 when the standard was revised? From the ISO 9001 Quality System Standard, paragraph 4.1.2.3—Management Representative:

*"The supplier's management with executive responsibility shall **appoint a member of the supplier's own management....**"*

Here is one last convincer. Within the organization's structuring of their EMS management, within procedures or a controlled organization chart, does this nonmanagement representative have defined and documented "official power" over management for

A. ensuring the effective implementation of the EMS?

B. reporting to management on the status for continuous improvement?

To summarize, the author has chosen to interpret this requirement as meaning a member of management will be the representative. The EMS auditor should verify this. Of course, for one last official interpretation of this requirement, check with your registrar.

The EMS auditor must also verify the roles, responsibilities, and authorities delegated to the management representative. In turn, a verification must be made to ensure that these defined and documented roles, responsibilities, and authorities are actually what the representative is doing. Included in this information will be how the representative reports to management on the performance of the EMS. By verifying this reporting process, this information on how this function may be performed may also become important information to consider when auditing 4.6-Management Review. Again, we see the interaction from one requirement of ISO 14001 to another.

Common System Nonconformances (CSN)

- **Roles, responsibilities, and authorities have not been effectively communicated to all relevant levels and functions within the organization's EMS.**

- **Resources essential to the implementation and control of the EMS have not been adequately provided for key operational personnel required to perform specific tasks which implement the EMS.**

- **Nonmanagement-level personnel have been granted the "official power" of being the management representative, where evidence in the audit has determined they do not have sufficient authority over management-level disciplines.**

EMS Auditor Checklist/Questionnaire

- How have *roles* been:

  A. defined

  B. documented

  C. communicated

  in order to facilitate effective environmental management?

- How have *responsibilities* been:

  A. defined

  B. documented

  C. communicated

  in order to facilitate effective environmental management?

- How have *authorities* been:

  A. defined

  B. documented

  C. communicated

  in order to facilitate effective environmental management?

- How has management provided resources essential to the:

A. implementation

B. control

of the EMS?

- What actions have taken place to provide the following resources:

A. human

B. specialized skills

C. technology

D. financial

- Have

A. defined

B. documented

C. communicated responsibilities and authority

been established for the management representative(s)?

- Are these roles, responsibilities, and authorities effectively implemented for ensuring that EMS requirements are

A. established

B. implemented

C. maintained

in accordance with the ISO 14001 standard?

- How does the management representative report on the performance of the EMS to top management for review and as a basis for improvement of the EMS?

ISO 14001, 4.4.2—Training, Awareness, and Competence

The organization shall identify training needs. It shall require that all personnel whose work may create a significant impact upon the environment have received appropriate training.

It shall establish and maintain procedures to make its employees or members at each relevant function and level aware of

(a)   the importance of conformance with the environmental policy and procedures and with the requirements of the environmental management system;

(b)   the significant environmental impacts, actual or potential, of their work activities and the environmental benefits of improved personal performance;

(c)   *their roles and responsibilities in achieving conformance with the environmental policy and procedures and with the requirements of the environmental management system, including emergency preparedness and response requirements; and*

(d)   *the potential consequences of departure from specified operating procedures.*

Personnel performing the tasks that can cause significant environmental impacts shall be competent on the basis of appropriate education, training, and/or experience.◆

Annex A.4.2—Training, Awareness, and Competence

The organization should establish and maintain procedures for identifying training needs. The organization should also require that contractors working on its behalf are able to demonstrate that their employees have the requisite training.

Management should determine the level of experience, competence, and training necessary to ensure the capability of personnel, especially those carrying out specialized environmental management functions.

...The organization shall identify training needs. The EMS auditor should keep in mind that an organization's needs must also be consistent with the requirements and the line item statement set forth within. Therefore, verification must be determined regarding this consistency. In addition, "identify" should be interpreted as "documented." The established procedures may include a listing of the specific training needs.◆

These needs shall require *all* personnel...All. The biggest little word in the English language. The EMS auditor must consider all management, supervision, and operational personnel whose work may create a significant impact upon the environment. The following may illustrate where the auditor may find examples of how the organization has determined the personnel-to-significant impact relationship:

- Lists of identified training needs
- Procedures for training
- Other EMS procedures for ISO 14001
- Organization charts

Appropriate training should be referred to as this consistent determination between the requirements and the identified training needs relevant to each level and function within the organization.

From our examples given earlier on the use of the UAC, this now becomes a critical data and information collection process. Now that we have acquired this information for other EMS audits, the EMS auditor may proceed to the activity responsible for maintaining training records to ensure objective evidence of the fact that the training has been carried out.

Specific to auditing this requirement, a comprehensive pre-audit review of the training procedures is conducted to ensure that they are established and documented. The EMS auditor will compare these procedures with the requirements of ISO 14001, 4.4.2, to ensure all elements **a through d,** including the final sentence, are incorporated into the procedures.

...shall be competent on the basis of appropriate education, training and/or experience.

The EMS auditor should consider several factors when determining "appropriate education and/or experience" is documented in lieu of the identified training needs. How relevant is past experience and education to the specific requirements listed in **a through d**? Many of these programmatic requirements are probably new to an organization, as is the overall EMS. But for personnel specifically working with regulatory compliance issues, some of their past experience and education may be very relevant to their knowledge of significant environmental impacts within their work activities. The terminology "and/or" leaves a great deal of flexibility for the organization in not only identifying their training needs, but also in how the organization chooses to document training in the form of records. The EMS auditor must, however, determine that the past experience/education is not only relevant, but is recognized and documented as training within the records.

An organization may also be considering integrating the existing training programs and procedures of their ISO 9001/2 Quality System with the training requirements of ISO 14001. For those of you with us today reading this book for insight into the development of your organization's EMS and supporting procedure—please be aware of the following.

For one, there is a significant difference in the requirements for training, awareness, and competence from one management system standard to the other. Second, there are significantly more elements to consider within the ISO 14001 standard. Finally, responsibilities, authorities, and personnel conducting the environmental training will greatly differ with respect to their own experience, knowledge, and education regarding the organization's key environmental issues.

A final comment on the overview of the ISO 14001 interpretation for audit needs to be addressed with regard to trainers. The EMS auditor must consider this important question when auditing the effectiveness of the environmental training processes. *How were you trained or qualified to be a trainer?* In qualifying such internal personnel, the EMS auditor should consider what appropriate education, training, and/or experience the instructors have to conduct this type of training. In verifying this, a sample of names of internal trainers/instructors should be documented. With this information, is there now evidence in the trainers'/instructors' records recognizing that they have received the documented education/training or have the experience to conduct such training? Food for thought…many a story evolved from ISO 9001/2 certification audits on this topic. And, strictly by coincidence, one of the key areas in question was the training of internal quality auditors. Evidence obtained throughout the audit gave every indication that the internal quality audit function was not competent, based on numerous findings in the certification audit, which were not detected by the internal quality audit function. This also brought to the forefront the competency of the internal quality auditing training process and that of its trainers/instructors. Further investigation disclosed that the instructors had no evidence of training relevant to the internal quality audit process themselves. In summary, the key to qualifying personnel is defining and documenting just what appropriate education, training, and experience is. To this end, the EMS auditor may now verify and understand what is appropriate and determine that these qualifying processes are carried out.

Common System Nonconformances (CSN)

- Education and/or experience is not included in training records.

- Training procedures do not address how the potential consequences of departure from specified operations are communicated to all applicable personnel.

- Management-level person responsible for the establishment, implementation, and control of EMS procedures has not documented training relevant to the requirements of ISO 14001 in training records.

- Office workers, administrative personnel, and/or contractors have not received appropriate emergency preparedness and response training, where applicable.

EMS Auditor Checklist/Questionnaire

- Gather all information and data from all other EMS audits conducted previously on training issues contained with the UAC (names, locations, identified training needs).
- Verify records are available and include evidence of training for each manager, supervisor, and key operational personnel (based on UAC sample).
- *Has the organization identified their training needs?*
- *Have all personnel whose work may create a significant impact upon the environment received appropriate training?*
- *Has the organization established and maintained procedures to make the organization's employees or members at each relevant function and level aware of:*
    - *(1) the importance of conformance with the environmental policy?*
    - *(2) the importance of conformance with environmental procedures?*
    - *(3) the requirements of the EMS (ISO 14001)?*
    - *(4) significant impacts, actual or potential, of their work activities?*
    - *(5) the environmental benefits of improved personal performance?*
    - *(6) their roles and responsibilities in achieving conformance with*
        - *A. the environmental policy?*
        - *B. environmental procedures?*
        - *C. requirements of the EMS (ISO 14001)?*
    - *(7) emergency preparedness and response requirements?*
    - *(8) the potential consequences of departure from specified operating procedures?*
- *Have trainers and/or instructors been qualified to conduct specialized environmental training?*

ISO 14001, 4.4.3—Communication

With regards to its environmental aspects and environmental management system, the organization shall establish and maintain procedures for

- *(a) internal communication between the various levels and functions of the organization; and*
- *(b) receiving, documenting, and responding to relevant communication from external interested parties.*

The organization shall consider processes for external communication on its significant environmental aspects and record its decision.◆

Annex A.4.3—Communication

Organizations should implement a procedure for receiving, documenting, and responding to relevant information and requests from interested parties. This procedure may include a dialogue with interested parties and consideration of their relevant concerns. In some circumstances, responses to interested parties' concerns may include relevant information about the environmental impacts associated with the organization's operations. These procedures should also address necessary communications with public authorities regarding emergency planning and other relevant issues.◆

Keeping the channels of communication open to internal and external parties is a vital element of the control and operation of the organization's EMS. Many forms of communication may be utilized, verbal or written, to transfer information relevant to the organizational environmental aspects and their EMS. For the EMS auditor, the pre-audit review of the organization's procedures for communication now becomes the specific internal standards against which to audit. The following is a listing of types of communications an organization may describe within their EMS and procedures for communication:

- changes to existing EMS policy and procedures
- changes to identified environmental aspects
- modifications to environmental management program(s) specific to the design of their objectives and targets
- changes or modifications to applicable environmental laws and legislation having a direct impact on a given function
- information regarding various levels and functions on environmental and EMS nonconformances
- information for management on the current status of the EMS operations through the management review process
- information to public authorities and other interested parties regarding emergency planning and response, including environmental regulatory compliance issues
- notification to internal and external parties regarding site management reviews, surveys, impact studies, and EMS and regulatory compliance audits and assessments

NOTE—The author wishes to point out these examples are not requirements the EMS auditor should look for, but typical examples of what an organization may choose to define as types of internal and external communications under the control of the EMS.

As we have seen from this listing, change or modifications to existing systems and operations could certainly be a vital element of communication. Notification and/or communication to effected "relevant functions and levels" regarding change is at the heart of an organization's controls for continuous improvement of their EMS.

And, by the way, just what does "relevant function and level" mean?

"Level" refers to span of control and authority, usually within a function. However, with the increasing prevalence of matrix organizations, level also can refer to lateral relationships that cut across two or more functions.

The phrase "relevant function and level" is intended to convey that ISO 14001 must be integrated throughout the organization. Broad deployment is essential in ensuring that the EMS is more than just another environmental, health, and safety (EHS) project.

Four elements in the standard make specific use of this phrase.

Section 4.3.3 requires documented environmental objectives and targets at each relevant function and level within the organization. Interpretation is straightforward—any function with activities that have actual or potential significant environmental impacts must develop objectives and targets that address those impacts. Similarly, any function with activities that are related to the commitments embodied in the organization's environmental policy must develop objectives and targets that ensure adherence to and implementation of those policy commitments.

The emphasis on activities within a function suggests that objectives and targets will affect all employees engaged in those activities. Successful achievement demands an understanding of and commitment to attaining objectives and targets from the function's junior members.

Section 4.3.4 requires designation of responsibility for achieving objectives and targets at each relevant function and level. Interpretation here is less clear-cut because of the manner in which the sentence has been constructed. Because "relevant function and level" modify "objectives and targets," some pundits believe that responsibility does not have to be pervasive. Discussions among members of the standard's drafting committee, however, suggest that the authors intended that responsibility for achieving objectives and targets must be driven down into the organization. If employees perceive such responsibility solely as a senior management obligation, the EMS will not be effective.

The third reference appears in **Section 4.4.2** which requires procedures to make employees at each relevant function and level aware of the EMS and their role in its successful implementation. This element has the broadest interpretation—employees throughout the organization who engage in any activities that are subject to the requirements of ISO 14001 must be made aware of that relationship and the organization's accompanying expectations.

The final use of the phrase occurs in **Section 4.4.3**, which stipulates procedures for internal communication between the various levels and functions of the organization and its environmental aspects and EMS. Seemingly straightforward, this element may pose the greatest challenge of the four. The requirement for communication procedures "between" levels and functions suggests the need for conduits that move information laterally among individuals at the same level, regardless of function, and "bottom up" from lower-level workers to mid-level employees and senior management down—communication methods are essential.[1]

...receiving, documenting, and responding to relevant external communications from interested parties.

"Interested parties" are individuals or groups concerned with or affected by the environmental performance of an organization. In determining and interpreting "relevant" external communications, any external party affected by the organization's environmental aspects and their EMS constitutes a relevant communication from an interested party. Receiving. Documenting. Responding to. These are three different things, are they not? The EMS auditor must remember to distinguish between the differences, especially in their relationship with environmental aspects and the organization's EMS.

...The organization shall consider processes for external communications on its significant environmental aspects and record its decision.

Consider. From Webster's Dictionary, "Think of the importance, implication, etc. of...."

There are several things for the EMS auditor to *consider*, by definition, when auditing this line item requirement.

First, if the interested party is affected by and has direct impact on the significant environmental aspect, the organization should do more than just consider looking into it. When reviewing the organization's record of its decision, the EMS auditor should ensure that this decision is documented, and that the appropriate designated internal personnel are, in fact, responsible for making such decisions. In addition, the decision that is documented must be a controlled document, right? The EMS auditor must also review to what extent the decision has been communicated to "relevant functions and levels" affected by such a decision.

The EMS audit function must always consider information and data compiled from EMS audits that may have a direct impact on other EMS audits. For example, does the decision made regarding the organization's significant environmental aspects have an impact on any existing environmental management program(s)? If so, how was this decision communicated to relevant levels and functions dealing with the specific objectives and targets. Also, was this significant environmental aspect identified as one in accordance with ISO 14001, 4.3.1 Environmental Aspects?

In summary, the EMS auditor must realize that an effective "communications" audit may take them into many different areas/functions within the organization. The EMS auditor must track the communication from its initial generation through the relevant levels and functions to determine if it does, in fact, get there and is acted upon, especially when the communication has a direct impact on systems or operational change and/or modification. Special attention should be paid to "verbal" communications internally to ensure that they are effectively received, understood, and acted upon. It must be determined by the EMS auditor to what extent established and maintained procedures describe the processes for verbal communications versus documented forms of communication.

Common System Nonconformances (CSN)

- **Procedures do not adequately describe the responsibilities and authorities of internal personnel required to:**

  A. receive,

  B. respond to, and/or

  C. document

  external communications from interested parties.

- **Documented internal communications were required by the procedure, but transmitted verbally.**

- **Records were not available at the time of the audit on external communications relating to the organization's significant environmental aspects.**

EMS Auditor Checklist/Questionnaire

- Has the organization established and maintained procedures for internal and external communications?
- How are the internal communications between various levels and functions of the organization carried out?
- How does the organization:
  A. receive
  B. document
  C. respond to
  relevant communications from external interested parties regarding its environmental aspects?
- How does the organization:
  A. receive
  B. document
  C. respond to
  relevant communications from external interested parties regarding its EMS?
- What processes has the organization considered for external communication on its significant environmental aspects?
- Have these recorded decisions been communicated to various levels and functions of the organization who have a direct impact on such decisions?

ISO 14001, 4.4.4—Environmental Management System Documentation

The organization shall establish and maintain information, in paper or electronic form, to

(a)   describe the core elements of the management system and their interaction; and

(b)   provide direction to related documentation.◆

Annex A.4.4—Environmental Documentation

The level of detail of the documentation should be sufficient to describe the core elements of the environmental management system and their interaction and provide direction on where to obtain more detailed information on the operation of specific parts of the environmental management system. This documentation may be integrated and shared with documentation of other systems implemented by the organization. It does not have to be in the form of a single manual.

Related documentation may include

(a) process information;

(b) organizational charts;

(c) internal standards and operational procedures; and

(d) site emergency plans.◆

There are several ways an organization may elect to establish and maintain information describing the core elements of the management system and their interaction, including:

- EMS Manual Table of Contents
- procedural matrix showing the interaction of management system elements
- flowcharts
- references within procedures to the interaction of other related management system procedures
- references of environmental documentation and records within the EMS or QMS procedures

This is, quite obviously, a relatively easy requirement against which to audit. It may be conducted during a document review of such information. The UAC should be considered on such information relevant to document control measures and responsibility and authorities for establishing and maintaining such information.

The EMS documentation audit, however, is of great importance to the EMS auditor. It may provide a clear insight on the relationship between the core elements of the EMS and the interactive processes which allow for the EMS to effectively operate within the organization's activities, products, and services.

For example, the author believes that one of the more important ways EMS requirements interact with each other is within the second core element of ISO 14001, under Planning. The following chart depicts the interaction of system requirements:

Requirement	Related Requirement	Interaction
4.3.1-Environmental Aspects	4.3.2-Legal and other Requirements	Identify/access to legal and other requirements applicable to environmental aspects
4.3.1-Environmental Aspects	4.3.3-Objectives and Targets	Establish/review objectives relating to legal and other requirements and significant environmental aspects
4.3.1-Environmental Aspects	4.3.4-Environmental Management Programme(s)	Establishing the means and timeframe for achieving objectives and targets relevant to identified significant environmental aspects
4.3.3-Objectives and Targets	4.3.4-Environmental Management Programme(s)	Establishing the means and timeframe for achieving objectives and targets relevant to identified significant environmental aspects

Strategic and effective planning requires the interaction of these requirements. Again, an organization may choose various ways of describing this interaction as referenced on the previous page.

Another key interaction between system requirements of ISO 14001 is the link between 4.5.4-EMS Audit and 4.5.2-Nonconformance, corrective and preventive action. It is interesting to note that this interaction now has a greater impact on the effective implementation of an organization's EMS. The requirements for EMS audit *do not* include a line item statement for conducting follow-up audits. Because of this oversight by the standard's developers, an EMS audit "finding" which has detected a nonconformance within a specific requirement of ISO 14001 is now interpreted as an "environmental nonconformance." EMS audit procedures should certainly contain a reference indicating, "Corrective and preventive action regarding EMS nonconformances shall be performed in accordance with Procedure Number XXXX, Nonconformance, Corrective and Preventive Action." Additional interaction of related system elements may include the following:

- 4.4.5-Document Control (regarding change)
- 4.5.3-Records (EMS audit, nonconformance reporting)
- 4.4.3-Communication (information provided to relevant functions)
- 4.4.2-Training, Awareness, and Competence (where applicable)

In summary, the audit of EMS documentation would be an excellent on-the-job training function for anyone involved in the EMS audit process, especially auditors. The author recommends, when the system is completely up and running, for EMS auditors to review, analyze, and understand this requirement, especially the core elements of the ISO 14001 standard and how the system interacts with itself specific to the design of the organization's EMS for a given facility. In addition, the author would further recommend that any newly-qualified EMS auditor evaluate this requirement before proceeding to other EMS requirements of ISO 14001.

Common System Nonconformances (CSN)

No CSNs specific to this requirement.

EMS Auditor Checklist/Questionnaire

- Has the organization established and maintained information
  A. describing the core elements of the management system?
  B. on the interaction of the core elements of the management system?
  C. on direction to related documentation?

ISO 14001, 4.4.5—Document Control

The organization shall establish and maintain procedures for controlling all documents required by this International Standard to ensure that

(a)   they can be located;

(b)   they are periodically reviewed, revised as necessary and approved for adequacy by authorized personnel;

(c)   the current versions of relevant documents are available at all locations where operations essential to the effective functioning of the system are performed;

(d)   obsolete documents are promptly removed from all points of issue and points of use or otherwise assured against unintended use; and

(e)   any obsolete documents retained for legal and/or knowledge preservation purposes are suitably identified.

Documentation shall be legible, dated (with dates of revision) and readily identifiable, maintained in an orderly manner and retained for a specified period. Procedures and responsibilities shall be established and maintained concerning the creation and modification of the various types of documents.◆

Annex A.4.5—Document Control

The intent of 4.4.5 is to ensure that organizations create and maintain documents in a manner sufficient to implement the environmental management system. However, the primary focus of organizations should be on the effective implementation of the environmental management system and on performance and not on a complex documentation control system.◆

Document control—unquestionably the "monster" of all management system requirements. The author wishes to note the disagreement with the final sentence of Annex A. While the primary focus of an organization's EMS should be on effective implementation and environmental performance, both are certainly in question without a complex documentation control system. How truly effective can an EMS be *without* the appropriate documents being available, reviewed and approved, controlled, current, and without methods for controlling obsolete documents? And, by the way, just what are environmental documents? The most important answer might actually be, "Whatever the organization says they are." But the real kicker here is the first line item statement of the requirement:

...*for controlling **all documents required by this standard**.* The following is a comprehensive listing of documents required by this standard. The EMS auditor should certainly consider using this list within their self-developed audit protocol and checklist when determining the level of controls applied to "required" documentation:

- Environmental policy
- All EMS procedures, where required
- Listing of (identified) environmental aspects
- Listing of (identified) significant environmental aspects
- Listing of (identified) legal and other requirements
- Environmental objectives and targets
- Environmental management programme(s)
- Roles, responsibilities, authorities (if covered on separate document and/or organization charts)
- EMS documentation (information, interaction, direction)
- Operational control plans (if documented)
- Operational control procedures and operating criteria
- Information to track environmental performance

- Management review documentation
- Flowcharts

Additional types of documentation required for reporting and/or recording will fall under the requirements of 4.5.3-Records. With respect to redundancy, the author draws your attention back to our ISO 14001, interpretation for audit of 4.3.2-Legal and other requirements. This may be an appropriate time for you to revisit this interpretation on document control of legal and other requirements.

As previously stated, establishing and maintaining procedures for controlling *all* documents certainly requires an understanding of just what "required" documents are. From the listing of environmental documents, we now must refer back to the UAC.

In preparing for any EMS audit, the effective use of the UAC relies on the EMS auditor's planning, preparation, and knowledge of environmental documentation relevant to any given audit. The author strongly recommends the development of a matrix covering the following elements:

- Type of documentation
- Document number
- Current revision and effective dates
- All locations documentation may be used
- Responsibility for control

The following example illustrates the use and development of the matrix:

Type	Doc No.	Rev	Date	Locations	Responsibility
EMS Audit Procedure	EMS 454	B	10/26/96	EM/QA	Director of Environmental Management, QA Manager
Emergency Preparedness and response procedure	EMS 447	C	5/25/96	ES&H Manufacturing	Director—ES&H Manufacturing Manager

By creating an EMS audit matrix to use with the UAC, this allows the auditor to know what documentation is required in what areas to determine the level of control by investigative questions within the UAC. For legal and other requirements, remember, there is already a requirement to identify them. By obtaining this listing, much of the work is already done for you. Through this listing, we may know better as to which activities are required to have access to them or how access may be obtained through a delegated organization. Each independent EMS audit now creates the ability to collect a sample number of relevant document numbers, their current status, and location to be used in the document control audit. Now the EMS auditor may proceed directly to the areas/locations where these documents are controlled to determine:

- approval for adequacy by authorized personnel
- obsolete document control
- identification of obsolete documents retained for legal and/or knowledge preservation purposes
- legibility and ready identification
- verification of retention times
- verification of procedures for responsibility concerning the creation and modification (change control) of various types of documents

From the listing of documents required by this standard, there are certain types of documentation well-suited for our document and data collection process. A significant percentage of documents relating to the EMS may, in the EMS auditor's view, come under question or suspicion. For instance, you are reviewing operational procedures regarding the handling of hazardous materials. The procedure being reviewed has markup changes (in red pen) on the document. This is a perfect example of the kind of document to record the document name, procedure number, current revision and dates, review and approval, and location. Further investigation at a later time during the EMS audit for document control will verify the current status of that procedure and the change control methods required by formal procedure for controlling all documents. Other types of documents in question or under suspicion may include:

- not available at the time of the audit
- posted "notes"
- old, dusty manuals in operational areas containing documents relating to the EMS
- old procedures stuffed in drawers and under desks related to the EMS
- documents with no review/approval status
- documents over three years old with no recent changes
- illegible documents
- most flowcharts

Flowcharts. Ah, yes. Let us discuss for a moment what a flowchart is and what a flowchart is not. A flowchart is an excellent type of "guide" to be used for training purposes. It is also well-suited to give an overview of the various operational processes in a given area. But is a flowchart a *procedure*? The EMS auditor should certainly take into consideration the use of the Flowchart Assessment Checklist referenced below:

FLOWCHART ASSESSMENT CHECKLIST

Q: Is the flowchart a controlled document?

Q: Is the formal procedure readily available?

Q: To what extent does the operator/worker follow the flowchart exactly as designed?

Q: Does the flowchart indicate HOW the task will be accomplished?

Q: Does the flowchart indicate key roles, responsibilities, and authorities?

Q: Does the flowchart reference applicable forms/records for documenting objective evidence?

The author wishes to relay to you what a flowchart *is not*: IT IS NOT AN EFFECTIVE PROCEDURE.

Oops! I can hear the rumbling from here...

Think about it. *Really* think about it.

Has anyone out there ever seen a flowchart that could stand up to the Flowchart Assessment Checklist (FAC)? And, by the way, where do flowcharts come from? Management. Engineering. These individuals who develop flowcharts write them specifically how *they* think the process or task should be accomplished.

All those squares, diamonds, circles!

Question number three on the FAC is certainly an interesting one. The EMS auditor should carefully follow operators, workers, supervision, and/or management to see that the flowchart is being implemented as designed.

The author certainly does not have any intention of offending anyone with us today in disagreement. *If* the flowchart (which is a controlled document, right?) is found to be an effective level of guidance or an effectively implemented procedure, after using the FAC, then hats off to the developer.

One last comment on flowcharts—in fifteen years of quality- and environmentally-related audits and assessments conducted by the author, *not one* of these audits has yet to produce a flowchart that was in conformance with document control requirements as indicated in the FAC—but there is always a first time! Let us move on.

Let us fast-forward for a moment to the requirements under 4.5.4-EMS audits. The audit program, including any schedule, shall be based on the environmental importance of the activity. Since a great deal of the EMS is operating under the specific guidance and control of required documentation, consideration should be given to the "environmental importance" of the document control audit. The author strongly recommends, through scheduling, conducting more frequent document control audits.

For example, let us say an organization has audited the required elements of the EMS under the core element of ISO 14001, 4.3-Planning. They are:

4.3.1-Environmental aspects

4.3.2-Legal and other requirements

4.3.3-Objectives and targets

4.3.4-Environmental management programme(s)

The sample of documents, data, and information gathered from these EMS audits, through use of the UAC, is now used to conduct an EMS audit for document control. This same philosophy and approach may also be used following the EMS audits of all requirements within:

4.4-Implementation and Operation

4.5-Checking and Corrective Action

This process for using the UAC is especially effective early in the life of an organization's EMS. Particularly prior to the third-party pre-assessment and/or certification audit conducted by the accredited registrar. Remember, an organization may always increase or reduce the number of audits to a given requirement based on (1) the environmental importance of the activity and (2) the results of previous audits.

An EMS audit program and auditor must keep in mind that an organization may choose to integrate existing Quality System document control procedures with that of their EMS requirements for document control. Should this be the case, document control procedures should be reviewed to determine the following:

- Responsibilities and authorities for EMS documentation
- If the preexisting Quality procedures meet all requirements of ISO 14001, 4.4.5-Document control

- If there is a master list of what specific EMS documentation that will be controlled by this procedure
- When a master list exists, if it contains all EMS documentation required by this standard

It is very important for the EMS auditor to remember that this audit is specifically tailored to determine the extent of controls applied to environmental documentation, not quality-related documents.

In summary, the author believes that the concepts, methodologies, and techniques expressed to this point are relatively new to auditors and management. This approach, utilizing the UAC as designed, has been proven, over many years, to be extremely cost effective in allowing the EMS or QMS audit program(s) to look much deeper into the system, while using less valuable time of auditors and auditees. By using the UAC by design, the EMS audit program may reduce the time it takes to conduct in-depth, effective audits by 50% in the following areas:

4.2-Environmental policy

4.4.1-Structure and responsibility

4.4.2-Training, awareness, and competence

4.4.5-Document control

4.5.2-Nonconformance, corrective and preventive action

4.5.3-Records

NOTE: The ISO 14001 interpretation for audit will discuss an effective impact the document, data, and information collection processes of the UAC will have on auditing the requirements of ISO 14001, 4.5.2-Nonconformance, corrective and preventive action. In addition, the UAC may also be equally effective for QMS audits to ISO 9001/2.

Common System Nonconformances (CSN)

- **EMS documentation is not available at the time of the audit at all locations where operations essential to the effective functioning of the system are performed.**
- **Obsolete EMS documentation is discovered in operational areas.**
- **EMS documentation contains no evidence of being reviewed, revised, and approved for adequacy by authorized personnel.**
- **Change control methods, as required by procedures, are not consistently carried out as designed.**

- **Flowcharts (refer to any one or all of the questions as indicated on the FAC).**

EMS Auditor Checklist/Questionnaire

The following checklist/questionnaire is based on the use of the UAC. The EMS auditor gathers together the information, document numbers, and data compiled from the following EMS audits:

- 4.3.1-Environmental aspects
- 4.3.2-Legal and other requirements
- 4.3.3-Objectives and targets
- 4.3.4-Environmental management programme(s)

Per the UAC, a random sample of data collected has gathered 12–16 documented examples of various environmental documents used within each of these requirements. The EMS auditor, through additional planning and preparation, should proceed to the specific areas for conducting the EMS document control audit.

NOTE—Be sure to include the responsibility and authority for document control from the procedures in your audit plan before beginning.

- *Are document control procedures readily available in all locations?*
- *Has the*

  (1) review

  (2) revision, as necessary

  (3) approval for adequacy

  of the collected sample of document numbers been performed by authorized personnel in accordance with the document control procedures?
- *From the collected sample, is the current revision of these documents the same as the status of documents found at that location?*
- *From the information obtained from above, are obsolete documents promptly removed from*

  (1) all points of issue?

  (2) all points of use?

  (3) possible unintended use?
- *Are obsolete documents retained for legal and/or knowledge preservation purposes identified? (Sample 2 or 3 in each location, where applicable)*

- *From the collected samples, was documentation*
  (1) legible?
  (2) dated (with dates of revision)?
  (3) readily identifiable?
  (4) maintained in an orderly manner?
- *From the collected samples, has the documentation been retained for a period specified in the document control procedures?*
- *Have responsibilities for the creation of the collected samples of documents been conducted in accordance with the procedures?*
- *Have responsibilities for the modification of applicable collected samples been carried by authorized personnel as indicated in the procedures?*

ISO 14001, 4.4.6—Operational Control

The organization shall identify those operations and activities that are associated with the identified significant environmental aspects in line with its policy, objectives, and targets. The organization shall plan these activities, including maintenance, in order to ensure that they are carried out under specified conditions by

(a) establishing and maintaining documented procedures to cover situations where their absence could lead to deviations from the environmental policy and the objectives and targets;

(b) stipulating operating criteria in the procedure;

(c) establishing and maintaining procedures related to the identifiable significant environmental aspects of goods and services used by the organization and communicating relevant procedures and requirements to suppliers and contractors.◆

Annex A.4.6—Operational control-text may be included in a future revision.◆

The classic organization concern rears its ugly head with respect to this requirement: how many are too many procedures, and how few are not enough to comply with this requirement? For the EMS auditor, our first objective is to *identify* the significant environmental aspects as defined within the objectives and targets. Second, *identify* those operations and activities that have an association with the significant environmental aspects. This information may be found contained within the defined objectives and targets. Third, the EMS auditor must determine how the organization shall plan these activities. This plan may be contained within the defined objectives and targets and/or documented procedures, giving clear guidance on how personnel will perform specific tasks in line with the policy and achieving objectives and targets.

The EMS audit should be carefully planned in order to sample various operational areas within the organization. Operational procedures should be reviewed and available at each relevant function and level of the organization dealing with specific, defined objectives and targets within the environmental management programmme(s).

...where their absence could lead to deviations...

The EMS auditor must use sound judgement when determining whether or not an activity *must* have documented procedures for a given operation. If the operation can perform tasks consistent with the means and time frame established for achieving objectives and targets *without* documented procedures, then hats off to that activity. If deviations are discovered in operational areas dealing with objectives and targets which do not contain documented procedures, then we have an EMS nonconformance. The appropriate corrective and preventive measures, per this requirement, would be to develop documented procedures for these operations. The absence (documented procedures) *did* lead to deviations.

The documented procedures should be verified by the EMS auditor as being consistent with the performance of carrying out the tasks set forth in the procedure. *Operating criteria* may define how these tasks shall be performed. In addition, the EMS auditor must always verify that these procedures are "readily available" to operational personnel required to use them.

"Oh, the operation is relatively easy. I have had extensive training in the procedures already, but they're over in the file cabinet in case you need to see them," says the operator.

"O.K., great. Could you bring me the procedures so that I can take a look at them," says the EMS auditor.

This is a perfect scenario, through your pre-designed audit plan, to verify that the operator is following the procedure by design. Often, when this occurs, the EMS auditor may discover that the operator is not following the procedure by design.

One of the most important characteristics of an auditor is to effectively determine how personnel regularly perform operations in accordance with procedures. When the EMS auditor discovers a deviation from the procedure, DO NOT SAY ANYTHING!

Let the operator continue to perform the task to determine if this deviation may have an impact on other specified operations. Throughout the review to this procedure, several other deviations have been discovered. Then, the EMS auditor may question, "I observed these specific tasks were performed this way, but the procedures indicate a different method." Let the operator go to the point of no return.

Remember, it is not our place, responsibility, or authority to correct the operator performing these tasks. We are to verify they are being performed in accordance with the documented procedures.

The operator then may say, "Oh, yeah. One of our environmental engineers came by last week and said to do it this way. It was determined this way is more efficient and will supply more information about this process. I assure you we're doing it the correct way."

"Correctness means nothing in auditing, only difference."

We now have an EMS nonconformance against the specific procedure. The EMS auditor also has another concern. There was a lack of adequate change control relevant to this procedure. Another perfect scenario for our document, data, and information collection process for the UAC. By documenting this information, i.e., procedure number, current revision, location, subject (type of operation), and the specific objective evidence regarding the nonconformance, this information may later be used in the document control audit to determine if change control measures were carried out in accordance with the document control procedures.

Another common concern the EMS auditor may have deals with training. Was, in fact, the operator adequately trained to the procedure? For this scenario, the author wishes to point you in another direction. Was the *environmental engineer* or *supervisor* or *management*-level personnel trained in the procedural requirements for change control? Per the UAC, several names may be recorded to determine to what extent responsible parties were adequately trained per this requirement.

It is *critical* for the EMS auditor to remember this: all individuals in question may have had extensive training in these matters. Do not assume anything in auditing. The EMS nonconformance will be followed up to determine the corrective and preventive measures applied. The investigation into the cause of the nonconformance should lead to a modification of the procedure, with required change control measures applied, to "close-out" this EMS nonconformance. Training may not be an issue.

The EMS auditor should also investigate how and to what extent the organization has planned for dealing with operations and activities associated with the identified significant aspects, and how these planned activities are maintained. This "plan" may be contained within the defined objectives for dealing with significant environmental aspects. This is another example of the interaction within specified requirements of ISO 14001, between objectives and targets and the required operational controls of planning activities and operations relevant to their environmental policy, objectives, and targets.

...procedures related to the identifiable significant environmental aspects of goods and services...

The EMS auditor must review documented objectives and targets to determine to what extent products and services are included in a given objective. Each relevant function and level in the organization has already been determined within the objectives. By determining the "functions and levels" involved with goods and services, the EMS auditor may now verify established and maintained procedures are effectively implemented. In addition, how have these procedures and requirements been communicated to suppliers and contractors who have a significant impact on the overall achievement of objectives and targets. To this end, the EMS auditor must consider questioning supplier and contracted personnel to determine if the procedures and requirements have been:

A. communicated

B. understood

C. effectively implemented to ensure the environmental policy, objectives, and targets are carried out as designed regarding the organization's significant environmental aspects.

In summary, the EMS audit program, function, and auditors must keep in mind that you can only "sample" various operational controls in one EMS audit. Due to the likely possibility that there may be extensive operational controls in various activities, the EMS audit function should consider more frequent audits for this requirement. In addition, consideration should be given to changing the operational controls and areas audited with each additional EMS audit performed to this requirement. Similar to document control, the EMS audit program may elect to conduct operational control audits at a three to one ratio to other EMS audits. This is, quite obviously, a critical process for determining the overall environmental performance of an organization. More frequent audits to this requirement may lead to a higher level of visibility to management for fulfilling their dedication and commitment to continuous improvement, conformance to the EMS, and improved environmental performance.

Common System Nonconformances (CSN)

- Operational procedures for maintenance are not available for operations affecting the organization's significant environmental aspects.

- Operational procedures are not readily available at the time of the audit.

- Communications on relevant procedures and requirements to suppliers and/or contractors could not be verified.

- Flowcharts used as operational procedures do not allow operators to perform all tasks as required by formal documented procedures.

EMS Auditors Checklist/Questionnaire

- *How has the organization identified operations and activities with the identified significant environmental aspects?*

- *Are these operations and activities consistent with the environmental policy, objectives, and targets?*

- *What steps has the organization taken to plan these identified operations and activities?*

- *Are the planned operations and activities being maintained by the organization?*

- *Have documented procedures been established and maintained for operations and activities dealing with identified significant environmental aspects (where applicable)?*

- *Are these documented procedures effectively implemented? (Sample 3 or 4)*

- *Are operating criteria stipulated in the procedures?*

- *Have documented procedures been established and maintained relating to the significant environmental aspects of*

  A. *goods*

  B. *services*

  used by the organization?

- *Are these procedures effectively implemented? (Sample 1 or 2 each)*

- *How has the organization communicated relevant*

  A. *procedures*

B.   *requirements*

to

A.   *suppliers?*

B.   *contractors?*

ISO 14001, 4.4.7—Emergency Preparedness and Response

The organization shall establish and maintain procedures to identify potential for and respond to accidents and emergency situations, and for preventing and mitigating the environmental impacts that may be associated with them.

The organization shall review and revise, where necessary, its emergency preparedness and response procedures, in particular, after the occurrence of accidents or emergency situations.

The organization shall also periodically test such procedures where practicable.◆

Annex A.4.7—Emergency Preparedness and Response-text may be included here in a future revision.◆

Many organizations may have already developed and implemented systems for emergency preparedness and response. There is, and always has been, a plethora of OSHA and environmentally related standards, regulations, and requirements relevant to this requirement. As we have seen to this point, especially from the EMS auditors checklist/questionnaire, it is imperative to conduct thorough EMS investigations and audits to the specific line item statements of each requirement. However, an organization may elect to develop and integrate these requirements with existing systems, taking the procedures well beyond the specific requirement of this standard. An EMS auditor has the responsibility to verify the effective implementation of an organization's procedures. In sampling the organization's system, the EMS auditor should remember that the requirements of this standard are an essential part of the EMS audit plan and audit process.

...procedures to identify the potential for...

An organization must consider what could possibly happen to create an accident or emergency situation. For the EMS auditor, a determination should be made as to how relevant various potential accidents and/or emergency situations are to specific operational areas within the facility. In identifying these situations within the procedures, is there a clear understanding of preparedness and response operations to be carried out by personnel in that area?

...and respond to accidents and emergency situations...

The EMS auditor, after determining personnel understand "potential" situations, should now verify their ability to respond to the situation in accordance with the documented procedures. A determination should be made as to how consistently personnel interpret what role they may play in the response process. For example, a fire has broken out in an organization's storage facility. Do the procedures address the following:

- Who is responsible for communicating this situation in the immediate area? Facility-wide?
- Who is responsible for coordinating the evacuation process?
- Who is responsible for simply getting out?
- Where do they *safely* go?
- Is emergency equipment readily available?
- Is there a clear understanding of what to do and how to use such equipment (fire extinguishers, sprinkler systems)?

An organization must interpret this term "respond to" in various ways to fit the specific accident or emergency situation. The EMS auditor should develop specific audit protocol and plans against the documented response procedures in several areas to ensure that the operational areas have the ability to effectively respond as designed. Per our example above, these may be very relevant questions for the EMS auditor to ask. There may, quite obviously, be many more. The EMS audit programme should also consider covering various areas of the facility not in immediate potential danger of the emergency situation or accident to determine what role they may play in the response process.

...and for preventing and mitigating the environmental impacts associated with them....

Prevention is at the heart of an organization's mind set for organizing, understanding, controlling, and responding to emergency situations. How preventive measures are applied and communicated to various operational areas, functions, and personnel may differ greatly from one situation to another. The EMS auditor must also recognize that an organization may reference this line item statement in procedures as follows:

"Preventing and mitigating the environmental impacts associated with accidents and emergency situations shall be performed in accordance with

Procedure Number 4.5.2, 'Nonconformance and corrective and preventive action.'"

Recognizing an accident or emergency situation as an environmental nonconformance may become commonplace with an organization's EMS. Also common are the requirements for dealing with nonconformances within 4.5.2. This may be a good example of the interaction between different requirements of this standard, as previously discussed in 4.4.4-EMS documentation. Under this condition, the EMS auditor may review the appropriate documentation for "**mitigating** any impacts caused and for initiating and completing corrective and **preventive** action," of course, this process may only be investigated and reviewed if there actually was an accident or emergency situation.

But what about accidents and/or emergency situations that have never occurred?

An organization, by requirement, will, "identify the potential for and respond to." The EMS auditor should review and verify how an organization has *planned* for prevention of an emergency situation and what measures are in place to prevent the environmental impacts associated with them.

From Webster's Dictionary, "mitigate" is defined as "lessen in severity." By definition, the EMS auditor should determine how the organization has identified the severity of the impact on the environment, and has, in turn, developed appropriate preventive controls to deal with the severity of the environmental impact encountered. This, too, may also be dealt with by the organization through their nonconformance, corrective, and preventive action procedures. And again, the EMS auditor must determine how the organization has determined this "lessening in severity" for accidents or emergency situations that have the potential of occurring.

...shall review and revise, where necessary, its emergency preparedness and response procedures...

"Where necessary" is a common "weasel" term that is commonly used in standards, regulations, and specifications. The EMS auditor should review the procedures and determine if the organization has identified where it is necessary. By requirement, verification should be made, *in particular, after the occurrence of accidents or emergency situations.*

The EMS auditor, while planning and preparing for this audit, should identify a past and reasonably recent emergency situation that has occurred at the organization's facility. From this obtained data and information, we may now more effectively investigate several of the requirements previously discussed relevant to that identified situation.

This brings up an interesting situation that may occur for the audit function, especially third-party audits.

This information may be confidential, proprietary information protected by attorney-client privilege. Because of potential regulatory issues, concerns, and/or potential liabilities connected to this situation, this information/data is unobtainable. Fine. The EMS audit function should certainly respect such a decision. However, the EMS auditor need also remember this: our objective is to verify conclusive, objective evidence that will clearly indicate a requirement of this standard has been effectively implemented by the organization.

If objective evidence is not readily available for review, investigation and evaluating against any requirement of this standard, the auditor has an EMS nonconformance. Period.

...The organization shall periodically test such procedures, where practicable....

They certainly jammed a few "weasel" terms into this requirement, did they not?

Periodically. The author truly believes you may see this documented in procedures: "The organization shall periodically test such procedures, where practical."

If this is a "stand alone" statement in a procedure, with no other information relevant to "how" they may test such procedures, "how often" is periodically, and where is it practical or not practical to test such procedures, this is an EMS nonconformance. The EMS auditor should determine how the organization has determined "how often" periodically is. From there, verification should be made that the organization has tested these procedures within a defined interval. Testing of these procedures may be conducted by an organization in any one or combination of the following:

- desktop exercises
- full-scale drills
- evacuation exercises
- emergency equipment testing
- emergency communications testing
- emergency preparedness and response training

It is, however, very important to note that the above are not specific requirements of this standard. They may, in fact, be methods the organization has developed and integrated into their procedures for "what" they may do to test such procedures. If this is the case, the EMS auditor will verify that testing methods have been conducted as designed by the organization within their procedures.

In summary, it is very important for the EMS auditor to recognize the fact that this requirement may be implemented in many different ways by an organization. Accidents and emergency situations can greatly differ from one facility's operations to another. So may the methods an organization has developed for preparing and responding to the unique situations that could potentially occur of their facility.

The EMS auditor must keep an open mind when auditing to this requirement, especially third-party auditors.

Common System Nonconformances (CSN)

- **The identification of potential accidents and emergency situations has not been identified in all operational activities where potential situations may occur.**
- **The organization has not determined the "lessening of severity," mitigation, of the environmental impacts associated with accidents and emergency situations.**
- **The emergency preparedness and response procedure has not been reviewed or revised as necessary, particularly after the occurrence of accidents or emergency situations.**
- **The organization has not supplied objective evidence of the testing of such procedures as designed.**

EMS Auditors Checklist/Questionnaire

- Has the organization established and maintained emergency preparedness and response procedures?
- Are these procedures readily available in all applicable operational areas?
- How has the organization identified the potential for accidents and emergency situations?
- How has the organization identified responding to accidents and emergency situations?

- What steps are taken for the prevention of environmental impacts that may be associated with accidents or emergency situations?
- What steps are taken for the mitigation of environmental impacts that may be associated with accidents or emergency situations?
- What steps are taken to
  A. review
  B. revise

  emergency preparedness and response procedures?
- How are the procedures
  A. reviewed
  B. revised

  after the occurrence of accidents or emergency situations?
- Upon verification, are the designated individual(s) having responsibility and authority for the review and revision of procedures identified and have they carried out this task?
- How has the organization periodically tested emergency preparedness and response procedures?

ISO 14001, 4.5—CHECKING AND CORRECTIVE ACTION

ISO 14001, 4.5.1—Monitoring and Measurement

The organization shall establish and maintain documented procedures to monitor and measure on a regular basis the key characteristic of its operations and activities that can have a significant impact on the environment. This shall include the recording of information to track performance, relevant operational controls, and conformance with the organization's objectives and targets.

Monitoring equipment shall be calibrated and maintained, and records of this process shall be retained according to the organization's procedures.

The organization shall establish and maintain a documented procedure for periodically evaluating compliance with relevant environmental legislation and regulations.◆

Annex A.5.1—Monitoring and Measurement-text may be included here in a future revision.◆

The primary components of an organization's EMS for monitoring, measuring, and evaluating environmental performance is at the heart of this requirement. Environmental Performance Indicators (EPI) are a means of evaluating and describing the environmental performance achieved by an organization. From ISO 14004, 4.2.5, Environmental objectives and targets, let us refresh our recollection of different types of measurable EPIs:

Progress towards an objective can generally be measured using environmental performance indicators such as:

- *quantity of raw material or energy used;*
- *quantity of emissions such as CO;*
- *waste produced per quantity of finished product;*
- *efficiency of material and energy use;*
- *number of environmental incidents/accidents;*
- *% waste recycled;*
- *% recycled material used in packaging;*
- *number of vehicle kilometers per unit of production;*
- *specific pollutant quantities (e.g., NO_x, SO_2, CO, HC, Pb, CFCs);*
- *investment in environmental protection;*
- *number of prosecutions; and,*
- *land area set aside for wildlife habitat.*

An integrated example of an EPI, as associated with objectives and targets, is shown below:

Objective: reduce energy required in manufacturing operations

Target: achieve 10% reduction of the previous year

Indicator: quantity of fuels and electricity per unit of production◆

The EMS auditor should ensure that the organization has established and maintained a system for measuring and monitoring against the organization's environmental objectives and targets in areas of management systems and operational processes.

...documented procedure to monitor and measure on a regular basis...

A determination should be made as to how the organization will "check and supervise" (monitor) key characteristics that have a significant impact on the environment. The measurement of these key characteristics and their results should be analyzed and used to determine areas of successful achievement of set objectives and targets and to identify activities requiring corrective action and improvement. For the EMS auditor, this information should be clearly stated within the documented procedures, including the identification of the organization's "key characteristics" to be monitored, measured, and evaluated.

The organization needs to also determine how often these key characteristics will be monitored, measured, and evaluated. "On a regular basis" may be determined relevant to the significance of impact on the environment. Objective, verifiable evidence should be evaluated by the auditor in determining the organization is performing these tasks at defined intervals specified in the documented procedures.

In preparation and planning for this audit, a review of the organization's identified significant environmental aspects, objectives, and targets is essential. In scheduling the EMS audit, consideration should be given to conducting this audit immediately after performing the EMS audits for ISO 14001, 4.3.3-Objectives and targets and 4.3.4-Environmental management programme(s). Much of the data and information collected may be effectively used by the EMS auditor in determining how the organization evaluates and analyzes environmental performance. The distinct interaction between these requirements allows the audit program to paint a clear picture for management to better track their efforts in performance of objectives, achieving set targets, and continuously improving the overall heart of the EMS.

The measurable output of the environmental performance data, and how it is used and analyzed, shall be recorded *to track performance* and verified by the EMS auditor in accordance with the documented procedures. The interaction of this performance, relevant to *operational controls, objectives, and targets* should be clearly identified *in the procedures and objectively verified during the audit.*

The EMS auditor should identify and develop a listing of various types of equipment used for monitoring environmental performance, their locations, intended uses, and responsibilities and authorities for calibration and maintenance of such equipment. It is very important to note that the line item requirements for calibration and maintenance in this standard are extremely limited. Audit protocol/checklists should be developed directly from the documented procedures to determine that the organization is performing these tasks as designed. An organization may elect to include such equipment within existing quality system procedures for the control of inspection, measuring, and test equipment. If their current quality system is in accordance with ISO 9001/2, the organization is subjecting this equipment to a significantly more stringent set of requirements. For the EMS auditor, a comprehensive audit on how the organization calibrates and maintains this equipment is performed against established quality system procedures. Remember, if nonconformances are detected against these procedures, this is an EMS nonconformance. It is important to understand, however, we are not doing a quality system (QS) audit. The QS procedure *is* the EMS procedure for calibration and maintenance of monitoring equipment. In addition, *records* of this process shall be verified by the auditor in accordance with the documented procedures.

The author wishes to draw your attention to probably the most significant flaw in this entire standard. There is, at present time, *no requirement for the calibration and maintenance of MEASURING EQUIPMENT*. Hello! The most important equipment an organization will utilize to record data and information for collecting, tracking, and analyzing environmental performance. The heart and soul of this EMS standard. The heart and soul of an organization's processes for continual improvement of the environmental performance of their operations and activities. And the measuring equipment is not required to be calibrated or maintained by this standard!

From ISO 14004, 4.4.2-Measuring and monitoring (ongoing performance):

...Appropriate processes should be in place to ensure the reliability of data, such as calibration of instruments, test equipment, and software and hardware sampling....◆

Notice how these "general guidelines on principles, systems and supporting techniques" do not physically say "measuring" equipment either? Hmm. But, regardless of what ISO 14004 says, we are not auditing to these guidelines, are we?

An organization should, however, keep in mind that several federal environmental regulations do require a calibration and maintenance system for environmental measuring equipment. The reliability of data produced by this equipment is vital to the integrity of the continuous improvement of environmental performance. It should also be noted that if an organization elects to calibrate and maintain environmental measuring equipment by procedure, it is fair game for the EMS auditor. And remember, if in doubt, always check with your registrar for their interpretation of this *nonrequirement*.

...documented procedure for periodically evaluating compliance with relevant environmental legislation and regulations.

Several steps must be taken to effectively audit this line item requirement. First, the EMS auditor should identify and document a sample of identified "legal and other requirements." These requirements are *relevant* because they are identified within the EMS. Second, a determination should be made as to whether these requirements are "periodically" evaluated for compliance. Finally, is the organization conducting these evaluations in accordance with the required EMS procedure? It is critically important for the EMS auditor to remember we are not performing a compliance-based audit. We are verifying conformance to a requirement within the EMS and a required procedure of 4.5.1-Monitoring and measurement.

In summary, the author strongly advises that a seasoned environmental person conduct this EMS audit. Without the appropriate technical background and knowledge in various environmental sciences, concepts, and performance issues relevant to the organization's activities, products, and services, it may be very difficult to effectively carry out this audit. In addition, the use of the same EMS auditor or the auditor who recently performed the objectives and targets and environmental management program(s) audit is recommended. In scheduling these audits and the interaction thereof, consideration should also be given to the review of different objectives, targets, and environmental performance programs within each consecutive "interactive audit" of these requirements.

Common System Nonconformances (CSN)

- **The monitoring and measuring of key characteristics are not performed within prescribed intervals (on a regular basis).**

- **The recording of information to track performance is not conducted within prescribed intervals (as defined).**

- **Records of calibration and maintenance of monitoring equipment were not available on all equipment monitoring key characteristics that can have a significant impact on the environment.**

- **All relevant environmental legislation and regulations were not periodically evaluated for compliance (based on sample).**

EMS Auditor Checklist/Questionnaire

- Has the organization established and maintained documented procedures to

  A. monitor

  B. measure

  key characteristics of its operations and activities that can have a significant impact on the environment?

- Do the documented procedures define intervals for

  A. monitoring

  B. measuring

  these key characteristics (on a regular basis)?

- Have the key characteristics for monitoring and measuring been identified?

- Are

  A. monitoring

  B. measuring

  activities directly associated with the organization's identified significant environmental aspects that can have a significant impact on the environment?

- Are records of monitoring and measuring activities available to track

  A. performance

  B. relevant operational controls

  C. conformance with the organization's objectives and targets?

- Has monitoring equipment been

  A. calibrated

  B. maintained

  according to the organization's procedures?

- Are there records available for the calibration and maintenance of monitoring equipment?

- Have these records been retained in accordance with the organization's procedures?

- What steps have been taken to identify monitoring equipment to be calibrated and maintained that are used to monitor key characteristics of its operations and activities which have a significant impact on the environment?

- Has the organization established and maintained a document procedure for evaluating compliance with relevant environmental legislation and regulations?
- Is there evidence that this procedure is being implemented?

ISO 14001, 4.5.2—Nonconformance and corrective and preventive action

The organization shall establish and maintain procedures for defining responsibility and authority for the handling and investigating nonconformance, taking action to mitigate any impacts caused and for initiating and completing corrective and preventive action.

Any corrective or preventive action taken to eliminate the causes of actual or potential nonconformances shall be appropriate to the magnitude of problems and commensurate with the environmental impact encountered.

The organization shall implement and record any changes in the documented procedures resulting from corrective and preventive action.◆

Annex A.5.2—Nonconformance and corrective and preventive action

In establishing and maintaining procedures for investigating and correcting nonconformance, the organization should include these basic elements:

- (a) identifying the cause of the nonconformance;
- (b) identifying and implementing the necessary corrective action;
- (c) implementing or modifying controls necessary to avoid the repetition of the nonconformance;
- (d) recording any changes in written procedures resulting from the corrective action.

Depending on the situation, this may be accomplished rapidly and with a minimum of formal planning, or it may be a more complex and long process. The associated documentation should be appropriate to the level of corrective action.◆

The first objective for the EMS audit function and auditor is to determine just what an environmental nonconformance is. It is fascinating to note that there is no guidance given whatsoever on defining types of environmental nonconformances within the EMS. There is also no specific requirement for an organization to define or list types of nonconformances within their facility's operations and EMS. This is yet another flaw in this standard that is bound to lead to some confusion in the future.

To assist in the elimination of possible confusion, let us discuss for a moment what an environmental nonconformance *is not*. An environmental nonconformance is not whatever an EMS auditor (including third-party registrars) says one is. For example, an EMS auditor may draw a conclusion that an "environmental nonconformance" has been discovered because the organization has not identified their waste streams as a *significant* environmental aspect. The first problem is, where does it say that a waste stream *must* be identified as a significant environmental aspect? Second, where does it say that this is an environmental nonconformance? EMS auditors must never read into the standard what is not there, nor should they form opinions with no factual line item statement of requirement on which to base the opinion.

For an organization, further elimination of this potential problem may occur if they simply identify types of environmental nonconformances relevant to the facility's operations and their EMS. The following are types of nonconformances which may be recognized by the organization:

- deviation from the environmental policy;
- deviation from environmental procedures within the EMS;
- deviation from specified elements of the environmental management programme(s), objectives, and targets;
- deviations from legislative and/or regulatory requirements;
- environmentally related accidents, spills, and/or emergency situations
- other related EMS deviations where no formal procedures are required, including
  - the management review process, and
  - the provision of resources essential to the implementation and control of the EMS.

These would appear to be the most common examples. However, these are not requirements. Prior to start-up of this EMS audit, the auditor must develop appropriate plans and protocol against the organization's recognized and/or identified environmental nonconformances and verify that the system and controls are in place as required.

...procedures for defining responsibility and authority for...

As previously discussed in 4.4.1-Structure and responsibility, these are two different things. Responsibility may be defined as "an individual or group performing the task," while authority may be a individual or group that "sees to it that it gets done." For example, a procedure may indicate that the Director of ES&H will receive, document, and respond to relevant communication from external interested parties. The EMS auditor, after a review of such documentation, has observed that an environmental engineer has actually documented and responded to this communication. Is this an EMS finding in the audit? Yes, indeed, it is. The EMS auditor should be aware of the potential "pass the buck" methods that often occur. If the procedure says that the Director of ES&H will, then that is what we look for and verify.

...handling and investigating nonconformance...

Again, these are two different things, are they not?

The EMS auditor should verify that there is a process in place for controlling, communicating, and dealing with the nonconformance (handling) and what steps will be taken to look into the significance of the impact this nonconformance has (investigating) on activities, products, or services. In addition, the procedures will clearly define these processes, and the EMS auditor will verify that the system is being implemented as designed.

...taking action to mitigate any impacts caused...

The EMS auditor will verify that the procedure addresses what steps will be taken to determine the severity of the impact on the environment and, in addition, addresses who is responsible and has the authority for taking such steps.

...initiating and completing corrective and preventive action....

Verification should be made that the procedure is implemented as designed regarding how the process begins and how the nonconformance is handled, investigated, reviewed, approved, and closed-out. Again, there may very well be several responsible parties involved in this process. A variety of different levels and functions within the organization may have decisions to make regarding how these processes are to be implemented within their effected activity.

CORRECTIVE ACTION. PREVENTIVE ACTION.

Ladies and gentlemen, these are also two *very* different things.

Probably the most common auditor mistake rears its ugly head relevant to these two requirements. Many "management system" auditors do not differentiate between the two. Many organizations do not differentiate between the two.

"Fix it." "Fighting fires." "Put a band-aid on it." "Crisis management."

Ever heard these terms before? Why? No, really, why?

The author truly believes that the primary reason is because many organizations do not do a very effective job of implementing PREVENTIVE ACTION.

First of all, corrective action deals with the necessary steps to be taken to eliminate the detected nonconformance. The EMS auditor should verify not only the steps taken, but that the nonconformance has, in fact, been corrected.

Preventive action is simply what steps need to be taken to ensure that this does not happen anymore! Or, because "stuff" happens, preventive action provides measures of control to ensure that a recurring nonconformance may be potentially eliminated before it ever happens. These are the primary bits of objective evidence that an EMS auditor must verify to ensure *effective* preventive measures are taken.

If your organization truly takes to heart the critical importance of preventive action, and follows through accordingly, then hats off to you. If not, then the author hopes that all EMS auditors will now take a closer look at how effective preventive actions are applied by the organization through this analysis. Let us move on.

...Any corrective or preventive action taken to eliminate the causes of actual and potential nonconformances...

Sit back. Relax. Here we go again.

The causes. Why did this happen? No, why did this really happen?

The following are the most common types of "symptoms" mistaken for causes:

(1) "We didn't have the time."

(2) "We didn't have the money."

(3) "We didn't have the resources (equipment or personnel)."

These are not the true reasons why a nonconformance has occurred. Effective root cause analysis should determine why management did not provide sufficient time, money, or resources to allow the process to be carried out as required. To this end, the EMS auditor should review and verify how the organization has determined what the true cause is and what appropriate steps are taken to correct and prevent these causes from recurring.

Oh, yes, and by the way....

The EMS auditor should remember that the "big three" symptoms are often types of verbal responses from the auditee when a nonconformance has been detected. This certainly raises another flag. Relevant to any EMS audit where these "excuses" pop up, there is also a concern over whether or not management is providing sufficient resources to effectively implement the EMS. However, these concerns must be verified. Data and information from this audit may include documenting examples of recurring nonconformances. This information, in turn, may be considered when conducting the EMS audit for 4.4.1-Structure and responsibility.

...appropriate to the magnitude of problems and commensurate with the environmental impact encountered....

The EMS auditor should review the corrective and preventive action process to ensure that it effectively deals with all pertinent functions and levels associated with this nonconformance. Does the organization consider prioritizing nonconformances based on the severity of the impact on the environment? Does it take into account all minor nonconformances observed? Are timely corrective and preventive measures taken, especially relevant to nonconformances having a potential impact on the safety and health of workers in operational areas? It is important to note that these questions may have direct bearing on how the organization's procedures are designed and not on the line item statements and requirements of this standard.

...The organization shall implement and record any changes in documented procedures resulting from corrective and preventive action....

The EMS auditor should sample *and* review two to three procedures with known changes resulting from corrective and preventive actions. Many of the questions from the UAC may apply here for document control. An investigation of the processes associated with these recently revised procedures should be conducted. At the physical location, again, document control questions from the UAC may apply. But most important, is there verifiable objective evidence that these changed procedures have been effectively implemented and have personnel been appropriately trained to perform the changed tasks accordingly? The EMS auditor should also confirm that the preventive measures are now an integrated part of the process.

Before leaving this requirement, the author needs to address one more flaw in this standard. First, let us take a look at ISO 9001 requirements for the control of nonconforming product:

...This control shall provide for identification, DOCUMENTATION, evaluation...of nonconforming product.

Well, you guessed it. There is no requirement to *document* the environmental nonconformance OR to document the investigating, handling, actions to mitigate any impacts, initiating, and completing of corrective and preventive actions.

All this requirement says is to establish and maintain procedures for defining **responsibility and authority**, and then it strings out this long sentence which is *directly related* to responsibilities and authorities. There is no line item statement of requirement to document this.

The author cannot believe it, either. OK, go ahead and check for yourself. I'll be here when you get back.

Amazing, is it not?

This really brings to light the critical importance of the design of the EMS procedures developed by the organization being audited. Chances are very good that an organization *will* document many of these processes to show objective evidence that the requirement has been fulfilled. If so, then the EMS auditor will review and verify that this documentation is completed as designed. But, again, DO NOT read the word "documented" into that first long, stringy sentence in this requirement. This standard has left it completely up to the organization, not up to the EMS auditor, to decide which of these processes will be documented.

Common System Nonconformances (CSN)

- **Responsibility and authorities are not defined for all activities included in the first long, stringy sentence of this requirement, including responsibilities and authorities for**
  A. **handling**
  B. **investigating**
  C. **taking action to mitigate any impacts caused**
  D. **initiating**
  E. **completing**
    (1) **corrective action**
    (2) **preventive action**
- **Preventive actions are not effectively taken to eliminate the causes of actual and potential nonconformances.**
- **The organization has not effectively implemented and/or recorded changes in documented procedures resulting from preventive actions.**

EMS Auditors Checklist/Questionnaire

- Has the organization established and maintained procedures for defining responsibility and authority for
  A. handling?
  B. investigating?
  C. taking action to mitigate any impacts caused?
  D. initiating?
  E. completing
     (1) corrective action?
     (2) preventive action?
- How does the organization take corrective action to eliminate the cause of
  A. actual
  B. potential
  nonconformances?
- How does the organization take preventive action to eliminate the cause of
  A. actual
  B. potential
  nonconformances?
- Are corrective actions taken appropriate to the magnitude of problems and commensurate with the environmental impact encountered?
- Are preventive actions taken appropriate to the magnitude of problems and commensurate with the environmental impact encountered?
- How has the organization
  A. implemented
  B. recorded
  any changes in the documented procedures resulting from corrective action?
- How has the organization
  A. implemented
  B. recorded
  any changes in the documented procedures resulting from preventive action?

ISO 14001, 4.5.3—Records

The organization shall establish and maintain procedures for the identification, maintenance, and disposition of environmental records. These records shall include training records and the results of audits and reviews.

Environmental records shall be legible, identifiable, and traceable to the activity, product, or service involved. Environmental records shall be stored and maintained in such a way that they are readily retrievable and protected against damage, deterioration, or loss. Their retention times shall be established and recorded.

Records shall be maintained, as appropriate to the system and to the organization, to demonstrate conformance to the requirements of this standard.◆

Annex A.5.3—Records

Procedures for the identification, maintenance, and disposition of records should focus on those records needed for the implementation and operation of the environmental management system and record the extent to which planned objectives and targets have been met.

Environmental records may include

(a) material on applicable environmental laws or other requirements;

(b) complaint records;

(c) training records;

(d) product process information;

(e) product information;

(f) inspection, maintenance, and calibration records;

(g) pertinent contractor and supplier information;

(h) incident reports;

(i) information on emergency preparedness and response;

(j) records of significant environmental impacts;

(k) audit results; and

(l) management reviews.

Proper account should be taken of confidential business information.◆

We journey back now to the UAC and the record collection process from previous EMS audits conducted by the EMS audit function. The EMS auditor(s) have collected information from various areas and locations of the facility that utilized EMS records. The following data has been collected:

- type of record and title
- record number (if applicable)
- location used

A sample of 20 to 30 records from various areas may now be taken to the appropriate location that maintains these records, including training records. Prior to start-up of this EMS audit, a desktop review and audit is conducted to verify:

...the organization has established and maintained procedures for the identification, maintenance, and disposition of environmental records.

By using the EMS auditors checklist/questionnaire in this section, the auditor will confirm that all line item statements are addressed in the procedures. In addition, a review of the methods of control contained within the procedures covering each line item of statement of requirement is conducted.

Following this review and desktop audit, the EMS auditor should develop a specific game plan for coordinating the records review and audit process with the specific locations where these records are retained. Proper scheduling and notification should be made to all applicable responsible persons. Within this audit game plan should be the specific retention times required for each record.

The author must address, at this time, another of the most common auditor mistakes in management system or compliance-based audits. Upon arrival to a given location where records are retained, the EMS auditor inquires, "Could you please get me three or four samples of this type of record for review?"

It seems like a fairly normal question, right? WRONG!

This is the worst technique an auditor could ever use. "Could you please get me...."

The auditee might be thinking, "Sure. I'd be happy to. In fact, we've got an entire file set aside for the auditor who asked that question."

All in which are probably the best, most complete records of their kind.

NOTE—In the author's experience with ISO 9001/2, there has been several occasions in which third-party registrars have used this technique—in certification audits!

The proper method for review and audit of records is depicted as follows:

"Where do you keep this type of record at this location?" inquires the auditor.

"We keep these records here in this file cabinet in my office," responds the auditee.

"Are these all the most recent records kept in this area?" asks the auditor.

"We retained them here for a period of three years, then they are discarded in accordance with our procedures," replies the auditee.

"Good. I'm going to *randomly* pull five or six samples from various files to review the records," states the auditor.

Randomly. Selected by the auditor. Throughout various files of records. During this selection process, the auditor should also keep in mind the established retention times for these records. In the above scenario, the records are kept for three years. The most common system nonconformance with records is that an organization has not "qualified" their retention times. In other words, the procedures in place for the EMS have only been in effect for one year, but their specified retention time for records is three years. This is, however, a very minor nonconformance within the records procedure.

The EMS auditor should pull one to two records for review to verify these retention times, with the majority pulled for review and audit within the last one to six months. The EMS auditor should certainly be more interested in how the system is in place at the time of the audit, with emphasis on determining that this objective evidence of the EMS (records) has been utilized consistently over a recent period of time (one to six months). The records are reviewed and audited against the following criteria:

A. legibility
B. identifiability
C. traceability to the activity, product, or service
D. ready retrievability
E. protection against damage, deterioration, or loss
F. retention times are verified and recorded

Additional audit questions will also arise for those records contained within computer systems and/or by electronic media. The EMS must ensure that these records are adequately reviewed, approved, and/or signed-off and that there are controls in place to protect against damage, deterioration, or loss, such as back up files.

In summary, the EMS audit program should take into consideration the types of records reviewed and audited in the initial audit and collect different types of records via the UAC for the next audit. This will allow for a more thorough investigation into the use, control, and maintenance of all EMS records.

Common System Nonconformances (CSN)

- **Retention times have not been qualified.**
- **A sample of records has indicated incomplete information contained therein.**
- **Records reviewed in the EMS audit indicate no objective evidence of the traceability to the activity, product, or service.**
- **Records are not maintained in the specific locations and areas as indicated by procedure.**

EMS Auditors Checklist/Questionnaire

- Collect sample data and information from previous EMS audits on EMS records (sample 10 to 20) using the UAC.

Desktop audit

- *Has the organization established and maintained procedures for the control of records, including*
  - A. *identification?*
  - B. *maintenance?*
  - C. *disposition?*
- *Do the procedures provide for methods of controlling records as follows:*
  - A. *legible?*
  - B. *traceable to the activity, product, or service?*
  - C. *readily retrievable?*
  - D. *protected against*
    - *(1) damage,*
    - *(2) deterioration, or*
    - *(3) loss?*
  - E. *retention times?*

Records Review and Audit

- Randomly select a sample of 5–6 of each record and ensure that the records are:

A. legible
B. traceable to the activity, product, or service
C. readily retrievable
D. protected against damage, deterioration, or loss
E. retention times can be qualified

ISO 14001, 4.5.4—*Environmental Management System Audit*

The organization shall establish and maintain (a) programme(s) and procedures for periodic environmental management system audits to be carried out, in order to

(a)   determine whether or not the environmental management system

    (1)   conforms to planned arrangements for environmental management including the requirements of this International Standard; and,

    (2)   has been properly implemented and maintained; and,

(b)   provide information on the results of audits to management.

The audit programme, including any schedule, shall be based on the environmental importance of the activity concerned and the results of previous audits. In order to be comprehensive, the audit procedures shall cover the audit scope, frequency and methodologies, as well as the responsibilities and requirements for conducting audits and reporting results.◆

Annex A.5.4—*Environmental Management System Audit*

The audit programme and procedures should cover:

(a) the activities and areas to be considered in audits;

(b) the frequency of audits;

(c) the responsibilities associated with managing and conducting audits;

(d) the communication of audit findings;

(e) auditor competence;

(f) how audits will be conducted.

Audits may be performed by personnel from within the organization or by external persons selected by the organization. In either case, the persons conducting the audit should be in a position to do so impartially and objectively.◆

The EMS audit of "our" requirements should certainly be conducted with a high level of impartiality and objectivity, even more so for internal EMS auditors. Unlike ISO 9001/2 requirements, this standard does not require personnel to be "independent" of those having direct responsibility for the activity being audited. By this, an organization may elect to conduct self-assessment activities performed by persons responsible for such activities. Again, being impartial and objective is essential for the internal EMS auditor to *determine whether or not the EMS has been properly implemented and maintained.*

...The organization shall establish and maintain (a) programme(s) and procedures for periodic environmental management system audits...

By definition, the program will plan the *method, list of subjects, and events* within the body of the procedures to assure the EMS conforms to *planned arrangements for environmental management including the requirements of this standard.* The EMS auditor must verify that the program is developed in such a way that all 17 primary requirements of ISO 14001 show objective evidence of the EMS audits carried out for each.

The program will *provide information on the results of audits to management.* Through this, the EMS auditor should review the programmatic procedures to verify how these results are communicated to management. They may be verbally transmitted at post-audit conferences and/or documented on EMS audit reports distributed directly to management of the affected activity.

EMS audits shall be scheduled. There are three primary verifications to be made with respect to the schedule. First, it must be verified that there is a document schedule, which must be a *controlled* document, at that. Second, it must be verified that there is objective evidence that audits are based on the environmental importance of the activity concerned. Verification should be made regarding how the "importance" is determined and referenced in the procedures. Finally, the results of previous audits will determine just how often the requirements of this standard may be audited.

...The audit procedures shall cover the audit scope...

The scope describes the extent and boundaries of the audit, including physical location, activities, and reporting. This may additionally include, by requirement, conducting the audit to "planned" arrangements. The EMS auditor should review a sample of audit plans developed in accordance with the procedure to ensure that the scope of the audit is defined and carried out as designed.

Frequency may best be verified within the EMS audit schedule. A determination should be made that the programme has, in fact, lived up to the design of the schedule with respect to how often each requirement of this standard will be audited.

The EMS auditor will verify the *methodologies* or a system of methods on how data and information are collected, reviewed, analyzed, and audited, including any checklists or protocol, to be used by the EMS auditor to effectively carry out the audit. *Responsibilities* within the audit programme must be defined and may include the following:

- EMS Audit Programme Management
- EMS Auditors
- Personnel of the audited activity (referenced)

The *requirements for conducting audits* may best be interpreted and verified as a combination of the methods, planning, scope, and specific audit protocol and checklists used to carry out each individual audit. *Reporting the results,* at a minimum, should include information on how the results of the audit are communicated and reported to personnel within the audited activity. This may additionally include a referencing to the management review process.

Do you notice anything missing from this requirement? How about the *follow-up*? At the present time, there is no requirement to effectively determine if the nonconformances discovered in an EMS audit have had the appropriate corrective and preventive actions applied.

This author believes that the standard writers and reviewers went completely comatose on this one. But remember, an EMS nonconformance discovered in an audit would fall under the requirements of 4.5.2-Nonconformance and corrective and preventive action, right?

Where does it say that in this standard? Is there a clear definition of what an EMS nonconformance is?

The EMS audit programme and auditors should certainly realize that an audit programme is not worth the paper it is written on if you do not carry out an effective follow-up. It is safe to say that many audit programs may include the following clause in their procedures: nonconformances discovered in the EMS audit will be followed up to assure corrective and preventive actions have been applied and that they are effectively implemented. Other programmes may reference the procedures for nonconformance and corrective and preventive action. That opens the door for the EMS auditor to review these processes against such procedures. It will be very interesting to see how organizations interpret the need for follow-up. It will be equally interesting to see what the third-party registrars will be looking for with respect to the "required" follow-up, and corrective and preventive action. Organizational personnel reading this book and making interpretations should certainly check with your registrar on this one.

An EMS auditor should recognize the fact that organizations may elect to integrate this requirement with their already established Quality System Internal Audit program requirements. Special attention should be given to the following:

- EMS programme and audit responsibilities and authorities
- EMS audit schedule based on environmental importance
- EMS auditor training and qualification
- EMS audit reporting (forms)
- Scope, frequency, and methods

There may be distinctly different forms, records, and processes within the procedures that allow for the EMS audit programme to effectively comply with the requirements of ISO 14001, 4.5.4. Our focus should certainly be on differentiating between the Quality audit versus the EMS audit, and how the organization may effectively implement the requirement of this standard.

There are a few advanced EMS audit techniques worth discussing, especially for second and/or third-party EMS audits. When conducting an entire systems audit to ISO 14001 (all requirements covered during one audit), the following sequence of operations is recommended (based on a two-person, three-day audit):

Step 1   When planning and preparing for the systems audit, schedule the audit for ISO 14001, 4.5.4-EMS audit, for *day three*.

Step 2   Following *day two*, compile a listing of all known EMS nonconformances discovered on *days one and two* between both auditors.

Step 3   On *day three*, during the audit of ISO 14001, 4.5.4-EMS audit, review past audit reports. Determine the extent to which the organization's audit programme detected the known nonconformances found during the systems audit.

Step 4   If similar known nonconformances were detected by the organization's audit programme, review, analyze, and determine whether or not the corrective and preventive actions taken were effective.

The EMS audit programme is the primary tool acting as the eyes and ears of management. The entire focus of the determination on effective environmental performance, continuous improvement, and conformance to this standard lies in the hands of the EMS audit programme and auditors. Therefore, the competency of the programme and auditors should be subject to the same level of review and oversight as other requirements, maybe more.

The author truly believes that this advanced audit technique is rarely used. But it is most effective. EMS auditors usually rely on a bunch of paperwork to tell them that corrective and preventive actions were taken and that they were effective. This advanced technique will verify how effective they really are. Hands on. Should an internal EMS audit function elect to use this technique, it should be performed by qualified auditors independent of the EMS audit programme.

Before leaving this requirement, the author wishes to address a rather unique interpretation once made by the RAB. Relevant to ISO 9001/2 certification audits, the RAB will occasionally perform surveillance audits on the accredited registrars. Verifications are made that the registrars are in conformance with accreditation system requirements for registrars and their own organization policies and procedures, and are effectively conducting the certification audits and interpreting the standard correctly.

The RAB wrote up nonconformances against several U.S. accredited registrars. The finding was, "Whoever audits the requirements of 4.17-Internal Quality Audits, may not audit any other requirement of the ISO 9001/2 standard due to potential bias." Officials at the RAB, as of April 1996, insist that they have never heard of this interpretation. A few U.S. registrars beg to differ.

This interpretation, however, does bring up an interesting point. Incorrect interpretations made by an auditor can reflect back on the audit organization. It may, in turn, have significant impact on the entire audit process.

Common System Nonconformances (CSN)

- **There is no objective evidence that EMS audits are scheduled based on the environmental importance of the activity and/or the results of previous audits.**
- **The EMS audit schedule does not reflect the actual dates and durations in which audits were performed.**
- **The EMS audit schedule has not been kept up-to-date for planning for future EMS audits.**
- **EMS audit programme and procedures do not address and/or have not implemented providing information on the results of the audits to management.**

EMS Auditor Checklist/Questionnaire

- *Has the organization established and maintained programme(s) and procedures for conducting periodic EMS audits?*
- *Do the programme and procedures:*
  - A. *conform to planned arrangements for environmental management?*
  - B. *include the requirements of this standard?*
  - C. *provide methods for determining whether or not the EMS has been properly implemented and maintained?*
  - D. *describe methods for providing information on the results of audits to management?*
- *Has the programme established and maintained an EMS audit schedule?*
- *How is the frequency of EMS audits based on*
  - A. *the environmental importance of the activity?*
  - B. *the results of previous audits?*
- *Do the EMS audit procedures cover the following:*
  - A. *scope*
  - B. *frequency*
  - C. *methods*
  - D. *responsibilities for conducting audits*
  - E. *requirements for conducting audits*
  - F. *reporting results*
- *How are corrective and preventive measures applied to EMS audit nonconformances (where applicable)?*

NOTE—Develop an audit checklist/questionnaire in accordance with the organization's EMS audit procedures for corrective and preventive action and follow-up (where applicable).

ISO 14001, 4.6—Management Review

The organization's top management shall, at intervals it determines, review the environmental management system to ensure its continuing suitability, adequacy, and effectiveness. The management review process shall ensure that the necessary information is collected to allow management to carry out this evaluation. This review shall be documented.

The management review shall address the possible need for changes to policy, objectives, and other elements of the environmental management system, in light of environmental management system audit results, changing circumstances, and the commitment to continued improvement.◆

Annex A.6—Management Review

In order to maintain continual improvement, suitability and effectiveness of the environmental management system, and thereby its performance, the organization's management should review and evaluate the environmental management system at defined intervals. The scope of the review should be comprehensive, though not all elements of an environmental management system need to be reviewed at once, and the review process may take place over a period of time.

The review of the policy, objectives, and procedures should be carried out by the level of management that defined them.

Reviews should include:

(a) results from audits;

(b) the extent to which objectives and targets have been met;

(c) the continuing suitability of the environmental management system in relation to changing conditions and information, and

(d) concerns among relevant interested parties.

Observations, conclusions and recommendations should be documented for necessary action.◆

Our final requirement of the ISO 14001 standard contains the same flaw as does the management review requirement of ISO 9001, 4.1.3. There is no requirement for management to establish and maintain *procedures* for the management review process.

This is undoubtedly one of the most important requirements of this standard. The EMS auditor must remember that objective evidence is essential in verifying that the requirements are effectively carried out by the organization's *top management*. First off, the EMS auditor must review and verify how often the organization's top management will review the EMS. Within ISO 9001/2 systems, this has classically been once a year. Most important in this review and verification is that there is objective evidence that the organization's defined top management has, in fact, conducted this review process.

Is a top management *designee* acceptable to sit in on this review? The requirement clearly states, "The organization's top management shall..." It does not say their designee. Under the defined requirements of 4.4.1-Structure and responsibility, the EMS auditor may determine who is top management. And remember, *this review shall be documented.* From there, it may be determined who, in fact, was involved in this management review process. If this documented review does not say who was there, does the EMS auditor have objective evidence that, "The organization's top management shall...?"

No, they do not.

Yes, this is an EMS audit nonconformance.

Suitability. Adequacy. Effectiveness.

By definition from Webster's Dictionary, *appropriate, sufficient results.*

The EMS auditor shall then verify that this review process ensures that the appropriate, sufficient results are used relevant to the organization's implementation of their EMS. These results are the *necessary information to collect to allow management to carry out this evaluation.*

The following, by requirement, is additional necessary information to collect and review:

- the possible need for changes;
- objectives and other elements of the EMS;
- audit results;
- changing circumstances; and
- commitment to continuous improvement.

Once again, this review shall be documented. The EMS auditor must ensure that documented objective evidence clearly shows that all of the above-mentioned necessary information to collect and review has been addressed in such documentation.

Common System Nonconformances (CSN)

- **Objective evidence of who within the organization's top management shall review the EMS is unavailable or incomplete.**

- **Management review documentation does not address changing circumstances affecting the EMS, nor how it was considered or reviewed.**

EMS Auditors Checklist/Questionnaire

- *Has the organization's top management participated in the review of the EMS to ensure its continuing suitability, adequacy, and effectiveness?*
- *Has this review been conducted at prescribed intervals?*
- *Has the management review process been documented?*
- *Does the documentation reference the necessary information to be collected and reviewed, including:*

   A. *the possible need for changes to policy?*

   B. *objectives and other elements of the EMS?*

   C. *EMS audit results?*

   D. *changing circumstances?*

   E. *commitment to continual improvement?*

In summarizing the ISO 14001 interpretation for audit, a few critical points must be revisited. First, the ***Annex A*** *is not auditable criteria.* It is strictly guidance for your consideration and review for better understanding the intent of this standard and the requirements.

There has also been much talk and many articles, publications, and training programs making a big deal about the differences between internal, second-party, and third-party management systems audits. Some of you with us today may have thought to yourselves, "Is this interpretation for internal auditors, or registrars, or what?" Let the author ask you something, "Are the requirements of this standard *different* for internal auditors or registrars?" No. Of course not. The following chapters of this book will illustrate different approaches throughout for preparation, planning, conducting the EMS audit, the auditee, reporting to management, EMS audit report writing, follow-up, and close-out used by various auditors, and audit functions. The interpretation remains the same for all, except if noted otherwise (which it is not). The author, as do you, encourages a universal interpretation of this standard with the intent to harmonize all engaged in its use into looking at it the same way. The real differences, obviously, are in an organization's procedures pertaining to their facility's unique products, activities, services, and related environmental issues of performance and regulation.

A final word, also on the use of the UAC—the idea here is not to reinvent the wheel, just improve it. Remember our given: We rarely have enough time to do everything we need to do in a management systems audit. This method efficiently produces more audit results using less auditor time. Give it a try.

The heart and soul of this standard is an organization's commitment to continuous improvement. To continuously improve the audit process will allow for increased visibility and awareness of how to continuously improve other programs, systems, and conformance to procedures of the EMS.

Endnotes

1. This article, "What Does 'Relevant Function and Level' Mean?" by Marylyn Block, originally was published in the March 1996 issue of the IESU, published by CEEM Information Services, 10521 Braddock Road, Fairfax, VA 22032, USA. Tel: +1-703-250-5900; fax: +1-703-250-4117; E-mail: <inquiry@ceem.com>; Website: <http; //www.tregistry.com/ceem/>. Reprinted with permission.

6

The Audited Organization

Introduction

This section may best be viewed as an introduction to the extensive study of the EMS audit planning and preparations chapter to follow. Prior to these operations, the EMS audit function must develop a complete understanding of the area, activity, and/or organization to be audited. There are many outside influences and attitudes, pro and con, which may have a significant effect on the audit and audit process. To this end, gaining a thorough knowledge about the organization and its key players and operational areas and activities is essential.

The majority of organizations will view the EMS audit as a necessary part of doing business. They realize that a cooperative effort will benefit them as well as act as a worthwhile process for the continuous improvement of their EMS. For the EMS auditor and audit function, much information is needed up front before starting the initial planning and preparations for the EMS audit. However, there are a variety of invisible factors integrated into the audit that may not be foreseen by the audit function. The following information will shed some light on these factors and will become some critically important audit techniques and considerations prior to the development of the EMS audit plans and execution of the audit.

Establishing Primary Contact

The audit function will establish a key contact person who will assist in providing the necessary information/documentation needed for the audit. In many cases, it may be the designated management representative. For internal audits, usually the management authority of the audited area activity or specific requirement of ISO 14001 becomes the contact person. Initial contact may vary in time frames from internal, second and third-party audits, as indicated in the following chart:

Internal EMS audit	15–30 days
Second-party (customer, corporate)	30–60 days
Third-party (independent, registrars)	60–120 days

Third-party audits, obviously, would be initiated by the audited organiza-tion. Depending on the independent or registration group conducting these audits, these may vary as well. Regardless of the time frames utilized, the EMS auditor and audit function must always keep in mind our given: *We rarely have enough time to do all we need to do in the EMS audit process.* In many cases, we should remember that it can take two or three times longer to plan and prepare for an EMS audit than it takes to conduct the audit. Because of this, ample time should be allowed to obtain all the necessary information needed to effectively plan and prepare for any type of EMS audit.

At the time of printing this book, the author is currently engaged in a rigorous effort of assisting in the development of several EMS programs at organizations in the United States. These initial processes include ISO 14000 awareness training, program development, initial environmental reviews, and gap analyses. In all cases to date, our contact person is the organization's designated "ISO-coordinator," ordinarily an ISO 9000 coordinator who landed the duties of initially setting up the efforts for development of their ISO 14001 EMS. In addition, these coordinators are not members of the organization's management and possess minimum technical experience in environmental issues at their facility. Interesting!

But an important lesson has been learned by the author in these efforts. All contact persons are very familiar with the "ISO certification process" already in place via their established certifications in ISO 9001/2. The avail-ability of management personnel overseeing environmental issues at their facilities has been immediate, with all cooperative efforts in place to obtain all of the appropriate information/documentation to successfully complete our contracted mission.

Many organizations will elect to integrate their existing quality manage-ment system with that of their EMS. This will allow for many of the organization's key personnel to engage in newly developed "systems" outside of their areas of expertise. But remember, many of these individuals did not have extensive backgrounds in "quality systems" before ISO 9000. The EMS audit function must not develop any preconceived notion about the organiza-tion because of who the key contact persons are. The audit itself will disclose appropriate levels of environmental training and awareness to effectively facilitate environmental management. Our contact person is merely the key person to assist us in obtaining what we need to effectively conduct the EMS audit.

Company Personalities

The organization, activity, or area being audited is normally comprised of management, supervision, and the work force. Within these different levels of responsibility and authorities, many different "personalities" surface that may have a significant impact on the audit process. The following is a listing of these personalities that may surface during the EMS audit:

- We're the greatest
- Yes and No'ers
- Motor mouth'ers
- Agree'ers
- Don't tell me about my shop
- How about lunch?
- The bleeding heart
- The "tell all" auditee
- The environmentalist/activist
- Takes everything personally
- What do I need to do to correct this?

☺WE'RE THE GREATEST☺

This is one of the most important techniques in EMS auditing. It is the reaction of personnel upon initial notification that an EMS nonconformance has been detected by the auditor. The EMS auditor must step back for a moment and observe how the activity responds to this information. The true dedication and commitment of the staff may surface at this moment. Is there visible evidence of the responsible personnel wanting to do the right thing?

The "we're the greatest" attitude is truly one of the most positive observations in the EMS audit. Many times personnel with appropriate responsibility and authority will immediately react to the situation in front of the auditor and start the corrective/preventive actions because of true concern over conformance issues to that procedure, area, or activity. Workers take real pride in doing the right thing and display a genuine interest in the audit process because they are truly interested in issues regarding conformance to requirements and continuous improvement. For the EMS auditor, this attitude is extremely noteworthy and should be considered for mention in the EMS audit reporting and post-audit conference processes.

The opposite reaction must also be discussed. When a real lack of interest surfaces by responsible personnel with a "who cares" reaction, this is of major concern to the EMS auditor. Keep in mind these individuals may, in fact, be the parties responsible for corrective and preventive actions to the cited EMS nonconformance. One of the constant factors in an EMS audit is the ability to determine true and effective implementation of the organization's environmental policy. Management, supervision, and the work force all play critical roles in driving the directives of the policy home.

One final comment in this area is with regard to individuals who challenge the interpretation being made by the EMS auditor. This topic will be discussed in greater detail in the following chapters, but it should be noted that this is a classic time waster for the EMS auditor. There is really nothing that can effectively be accomplished at this point during the audit. Engaging in extensive conversation on issues of interpretation is not a part of an effective EMS audit plan, checklist, and/or questionnaire. Consideration must be given to moving on and allowing the audited organization to formally deal with this situation through responding to any draft reports issued in this matter.

☺YES AND NO'ERS☺

These are individuals who will not give you a clear answer. The EMS auditor must determine, based on the specific line of questioning, if a "yes" or "no" answer is sufficient in detail to effectively determine conformance to a given requirement. Sometimes we may need to change our questioning approach to get a more detailed response to the operations being reviewed.

☺MOTORMOUTH'ERS☺

More common in the EMS internal audit function are the type of individuals who want to talk about everything except the information related to the EMS audit. They may be friends or colleagues with which you used to work. It may be necessary for them to update you on everything new that is going on, including the recent company picnics, their family, their new boat, the football game last night, and so on. The EMS auditor should politely indicate our "given," and that we would love to discuss these issues at a later time, but right now could you please show me the EMS procedures?

☺AGREE'ERS☺

Another extremely noteworthy personality to consider are responsible individuals who, after being informed of issues of nonconformance detected in the EMS audit, are completely aware of their existence. They may say, "Oh, I know. I've been aware of the fact we're not doing this for some time now, but with the shortage of manpower we have in this area, we just haven't been able to get around to doing that." The impact that this has is greatly increased regarding issues of regulatory compliance.

ISO 14001, section 4.3.1, states, "Management shall provide resources essential to the implementation and control of the EMS." Also noteworthy with this reaction is the "volunteered" information over and above the line of questioning. The EMS auditor must take note of these comments. A significant number of EMS "known" nonconformances may call for an in-depth investigation into these resources, or lack thereof, to determine what steps may be taken to allow for sufficient resources to effectively control and implement the EMS. This may be reviewed during the corrective and preventive action follow-up audit conducted by the EMS audit function.

☹DON'T TELL ME ABOUT MY SHOP☹

"I've been here for 30 years, long before this EMS stuff was ever required, and we've been doing just fine without it!" This attitude has appeared to greatly decline over the years, especially for and within facilities already certified to the ISO 9001/2 standards. However, the author truly believes that there is "one in every organization." Similar to the personalities who constantly challenge an EMS auditor's observations and findings, this attitude must be addressed. Chapter 12 on EMS reporting will discuss the methods and techniques for reporting and disclosing such information during an EMS audit.

☺HOW ABOUT LUNCH?☺

This approach is another of the classic time wasters. The organization may be sincere in showing their hospitality to various audit functions, especially second and third-party auditors. Being a gracious host can be an effective method for opening up channels of communication between auditors and auditees. It is also an effective method for keeping the auditors away from the areas to be audited. Consideration in this event must be given to how the EMS audit function has planned the audit, including allocating time for this activity.

☹THE BLEEDING HEART☹

This personality has surfaced several times in the author's experiences, especially during after-hours activities, away from the facility, such as dinner. After some time of general conversation, a "concerned individual" will feel the necessity to "confide" in you on several issues that may have a major impact on the facility's operations or their EMS. They may be truly dedicated individuals who may "volunteer" their deepest concerns about the facility and top management's attitude regarding why this cannot be done or that cannot be done. Very important for the EMS auditor is the fact that this information, such as it may be, must be verified and reviewed during the audit to assure objective evidence is obtained to validate the information given. Many times the author has "slightly adjusted" some of the areas to be audited or line of questioning to take into consideration this newly-discovered information.

Great discretion needs to be taken in following through with these verifications. The information reported on and disclosed as a result of these issues must be that of what was physically discovered during the EMS audit at the time of the audit. *NEVER* rely on verbal information as an absolute until it is formally reviewed and verified.

☺THE "TELL ALL" AUDITEE☺

This is another critically important technique in EMS auditing. Many times during just about any and every type of EMS audit, the auditee will "volunteer" information to the auditor. It goes something like, "I know this is the new procedure to cover these activities, but I'm sure it hasn't been distributed throughout the facility yet." Or, the common, "Well that's not exactly how we do it here." When these situations occur, the EMS auditor must be flexible in the line of questioning to pursue and investigate the status of this volunteered information. This may have an impact on your EMS audit or the audit being conducted by another EMS audit team member. Should a response from the auditee cause suspicion or concern, the EMS auditor should immediately question the information given appropriate to the requirements for that area or activity. In addition, detailed notes should be taken if the volunteered information affects another activity or area, and the notes should be passed along to the appropriate team member conducting that EMS audit for further investigation and review.

☹THE ENVIRONMENTALIST/ACTIVIST☹

These have been found to be somewhat rare personalities in an environmental audit, but have surfaced occasionally in the form of whistle-blowers. During investigation of environmental regulatory compliance, individuals who may be held liable for noncompliance issues "buckle under" to the pressures of what appears to be threatening situations to their jobs or relevant legal issues that they may be held accountable for. The author has been in this situation before, and it is not very comfortable, at that.

The environmental auditor or EMS auditor must obtain conclusive, objective evidence and tell it like it is regarding the observations and findings detected in an audit. An important factor here becomes the methods used for disclosure. Chapter 12 on EMS reporting will discuss appropriate methods for addressing the system nonconformances without pointing the finger at individuals who are responsible for issues within the system.

☹TAKES EVERYTHING PERSONALLY☹

The EMS auditor must be aware of the fact that nonconformances detected in the audit may create an abundance of additional effort and work for the audited organization. Occasionally, management-level personnel may take things personally and become nervous, agitated, disgruntled, and then all too commonly, "clam up." The latter refers to an approach that an auditee might take in not saying anything more to increase the severity or impact the audit may have on their activity.

The EMS audit function attacks the system, not individuals. Should situations arise where this uneasy feeling occurs, then the EMS auditor must remember, "we are only human." A straightforward, professional approach, showing no facial or verbal signs of disgust or concern, is certainly in order. Remember also, in the EMS audit, that chances are very good that you will find substantially more conformance than nonconformance. This certainly may be an opportune time to "accentuate the positive" and make mention of the good points of the audit.

☺WHAT DO I NEED TO DO TO CORRECT THIS?☺

Often during the audit, individuals of responsibility will ask the auditor this question. Many times the author has heard about the type of auditor who, at this point, would say, "That's not my job or position to comment." This is *not* the correct approach!

Auditors who are reluctant to give general recommendations in auditing usually lack a key characteristic of the audit process—they do not know how, nor does their management who enforces these rules.

An EMS auditor must not make a binding commitment to an organization regarding their efforts to provide effective corrective and preventive actions. The key to answering this common question, in actuality, is quite common:

> WHAT NEEDS TO BE ACCOMPLISHED,
>
> NOT HOW TO ACCOMPLISH IT.

Auditors are rarely in a position to tell anyone how to do anything. But the types of issues an organization needs to consider regarding information, requirements, and fundamental methods of control may be where an auditor can play a vital role in the processes of continuously improving and complying with their EMS.

Think about it.

Who is in a better position to provide comment and recommendations on WHAT needs to be accomplished than the auditor, who has extensively reviewed the requirements and the EMS procedures, and who may have audited systems before that showed ample evidence and examples of conformance?

One of the U.S. registrars working with ISO 9000 has taken an effective approach in this area. During pre-assessment or certification audits, system nonconformances may have been detected prior to the audit when procedures are not written in sufficient detail to cover all of the specific requirements. Their response back to the audited organization has been in the form of a question.

"What methods are used to document corrective actions taken to assure that they are effective?"

This approach gives no indication, or commitment thereto by the registrar, on how the organization may correct the detected nonconformance. But by answering this question within the appropriate procedures and through effective implementation, the cited nonconformance may be corrected.

This approach will be discussed in greater detail in Chapter 12 on EMS audit reporting.

The Escort

The escort is of great assistance to the EMS auditor. The escort is the key communication link with all facets of the facility's operations relevant to the area to be covered by the auditor. Prior to commencing with the EMS audit, notification must be given to the audited organization regarding supplying the EMS auditors with escorts. A very important point to remember about escorts—they may not be the person responsible for various operations and activities within the scope of the audit. In many cases, they are merely the individuals who can get you there to the appropriate areas, activities, and personnel necessary for carrying out the EMS audit.

Consideration must also be given to keeping each audit group as small as possible: the auditor, the escort, and individuals directly responsible for operations within the scope of the audit. When more individuals are with the group, little else may be accomplished. In fact, disruptions and slow downs are likely to occur.

The author believes many of you with us today are reading this book to gain not only advanced insight on the EMS audit, but to obtain a competitive advantage in the program and procedural development processes of your EMS. To that end, the author has always been quite puzzled by one simple question, "Why aren't more auditors escorts?"

Who else would you rather have escorting EMS auditors throughout the facility than a trained and qualified EMS internal auditor? They may assist in the facility's preparation for the audit, anticipate typical questions to be asked in a given area, and provide professional guidance on issues raised during the audit. The author's company, Quality Systems Development, has had a slogan for many years now:

> "THE BEST AUDITEE IS A QUALIFIED AUDITOR."

Did you ever wonder why audit trainers and course providers have never offered auditee training? Hmm, this gives me an idea.

Summary

In general, organizations and their key responsible personnel take great pride in their work. However, the many years of ruthless, unprofessional auditors have left some organizations with their guards up and on the defense during audits. This personality, like the ones mentioned in this chapter, can surface. In fact, there are still some organizations who review, approve, and discard the internal auditors' reports as a part of practice. Imagine that!

The EMS auditor must be flexible, professional, and extremely diplomatic throughout all phases of the audit. Being unintimidated is also a major plus, especially for internal auditors. We have as our primary goal the ability to be the eyes and ears of management and their efforts in continuous improvement and conformance with their EMS. Should personalities surface in the EMS audit, and they always do, we must stick to our guiding principles, disciplines, and audit plan to carry out the audit as designed.

7

EMS Audit Planning

Introduction

Planning and preparing for the EMS audit may be the fundamentally most important aspect of the entire audit process. Through this chapter, a common philosophy will be stressed. The extent to which an organization plans and prepares for each EMS audit is reusable data. That is, the audit function may initially develop extensive audit files, containing a plethora of information, documents, data, notes, and results of past audits to review against the next time around. In initial development of the EMS audit programme, an important consideration must be taken. It may take two to three times longer to plan and prepare for an EMS audit that it does to actually perform the audit.

The author believes that several management-level personnel just hit the roof with that one. The fact is that most organizations DO NOT allocate sufficient resources, time, and personnel to the audit planning processes. For the EMS auditor, heed the warning—if you are ever handed a checklist and asked to "go out there and see to it that we're doing this," my point has just been made. Answer some of the following questions if this situation sounds familiar:

- Did you, the EMS auditor, actually develop these checklists?
- Have you had ample time to review EMS procedures and determine key responsibilities and authorities for the audit?
- Have the procedures been reviewed against the applicable requirements prior to the audit?
- Have the checklists been reviewed against the requirements?
- Have changes to the EMS been taken into consideration prior to this audit, including new or updated procedures, operational activities, or personnel?

- Have results of past EMS audits and specific auditor notes been taken into consideration prior to execution of this EMS audit?
- Do the audit procedures and checklists cover the following
  - audit scope?
  - methodologies?
  - frequency (based on environmental performance of the activity and past audit results)?
  - requirements for conducting the audits?
  - methods for reporting results?

If your organization can answer "yes," to each of these questions for each and every EMS audit performed for all of the requirements of the ISO 14001 specification, then hats off to you. In your efforts of continuously improving the EMS audit programme, including the planning and preparations for the audit, let us now explore some of the methodologies and techniques for executing this process.

EMS Audit Questionnaire

An essential part of the EMS audit process is gaining as much knowledge and information about the audited organization as possible. The EMS audit questionnaire is designed to do just that. For second and third-party EMS auditors, this information becomes the type of reusable data that may be used to compare changing circumstances relating to future EMS audits at that facility. Especially relevant to key management personnel with various departmental responsibilities and authorities within the EMS, the questionnaire may also be used to request the appropriate EMS documentation required for review, analysis, and development of site-specific protocol for conducting the audit.

Requesting this information from the audited organization should take place 60–120 days prior to tentative pre-assessment/audit dates. A complete set of EMS documentation should be received prior to review and development of audit plans and protocol. Consideration should be given to installing a mechanism for ensuring all information/documentation is received, including follow-up request questionnaires for the additional data. Once all information is received, the EMS audit organization must now determine the following:

- knowledge and expertise of available audit personnel specific to the industry and environmental regulations, aspects/impacts, and processes of the audited organization;
- availability of audit personnel;

- coordination of scheduling team members for review, analysis, audit planning, and protocol development;
- number of auditors and audit-days required to conduct the audit; and
- selection of EMS lead auditor and team members.

The EMS audit questionnaire will be followed by a summary of key information and its use and intent for EMS audit programme personnel and auditors.

EMS Audit Questionnaire for ISO 14001			
1.0	**COMPANY NAME**	**Phone Number**	**Date**
1.1	**Address**		
1.2	**Products or Service**		
1.3	**Description of product(s)**		
2.0	**COMPANY BACKGROUND**		
2.1	**Type of ownership**		
2.2	**Length of time in business**		
2.3	**Number of plants**		
2.4	**Locations**		
2.5	**Describe hazardous material operations or areas**		
2.6	**List safety/health precautions/environmental awareness training required for entrance to any work areas**		
3.0	**SHOP CAPACITY**		
3.1	**Capacity currently utilized**		
3.2	**Number of buildings**		
3.3	**Number of shifts**		
Working hours:		**Office:**	**Shop:**

4.0	Government/state, federal, and/or local environmental regulatory agencies or bodies applicability			
5.0 KEY CONTACT PERSONNEL		Name	Title	Dept. No. of Employee
Top Management				
Purchasing				
Environmental Compliance				
Environmental Engineering				
Manufacturing/Production				
Quality Assurance				
Engineering				
Shipping/Receiving				
Material Control				
Material Storage				
Human Resources				
Supplier Control				
Marketing/Sales				
EMS Document Control				
EMS Audit Programme				
Safety/Health				

6.0	QUESTIONNAIRE: Please complete the following and return to [Audit Organization] within 15 days:
6.1	Is your organization currently YES ❑ NO ❑ ISO-9001/2 certified?
	If YES, Registrar: Date:
6.2	State the period of time allocated for design and develop ment of your organization's EMS to ISO 14001
6.3	State the effective date of implementation of:
	EMS Policy
	EMS Procedures
	Environmental Management Programme
6.4	Has the EMS been integrated with your preexisting ISO 9000 Quality Management System? (Where applicable) YES NO

6.4.1 If YES, please indicate existing programmes and/or procedures used for both the Quality and Environmental Management Systems:			
Procedure Number	**Title**	**Effective Date**	**Revision**

6.5	Tentative date for pre-assessment audit (if applicable)
6.6	Tentative date for certification audit (if applicable)
6.7	When did your organization begin conducting EMS internal audits?
6.8	Have all requirements of the ISO 14001 standard been audited to date?
6.9	When did your organization begin conducting periodic evaluations to relevant environmental legislation and requirements?
6.10	Are you currently conducting environmental performance evaluations (EPE)?
7.0	Upon completion of this questionnaire, please submit questionnaire and the following EMS documentation to [Audit Organization]:
7.1	❏ EMS policy
7.2	❏ Relevant EMS procedures
7.3	❏ Listing of controlled environmental documentation
7.4	❏ Listing of controlled environmental records
8.0	A formal cover letter containing the EMS audit agenda and confirmation of scheduled date(s) for EMS audit(s) and/or assessment(s) will be sent to your organization 30 days in advance of the event. If you require additional information, please contact our [Audit Organization] office. *Thank you.*
[Audit Organization Manager]	Date

Review

While much of the ISO 14001 questionnaire is self-explanatory, let us discuss some of the key elements of this document and the purpose of its use.

Line 2.5-Hazardous material operations

Prior to execution of any type of external EMS or environmental compliance audit, the need for this information is essential. The EMS auditor must consider and plan for, where necessary, the types of precautions, personal equipment, and safety requirements for entrance into any operational area. Long delays could occur during the EMS audit if these situations arise without prior knowledge of their existence. Line 2.6 (training) may also become necessary prior to entrance and must be incorporated into any pre-audit planning time schedules for conducting area-specific EMS audits.

Line 3.3-Number of shifts

This is probably one of the most overlooked aspects of any type of management system or compliance-based audit. The EMS auditor should determine what impact operations essential to the effective control and implementation of the organization's EMS exist on second and third shifts. Consideration should be given to evaluating how such systems and information are controlled and are made available to all personnel effected by the EMS. This very well could and should include an examination of significant operations taking place during these shifts.

Line 5.0-Key personnel

Policy, procedures, and organizational charts will play a major role in determining where the lines of roles, responsibility, and authority lie within the EMS. The EMS auditor may, however, go one step further. Knowing just who these individuals are should be predetermined in audit plans and protocol.

During a quality systems audit of a major defense contractor conducted by the author several years ago, the available data regarding personnel was compared to a previous questionnaire sent to the contractor several years earlier. It was immediately determined that a significant number of these key management-level personnel had changed—and so did the status of their quality system. This element of a questionnaire becomes the type of reusable data previously mentioned. The EMS auditor should consider exploring such areas where recent changes to key personnel have occurred and review the status of systems operations in these areas to determine what impact these changes have had on the effective implementation of the EMS.

Line 6.4-Integrated management systems procedures

It is quite possible that an EMS auditor will review and audit against quality management procedures in an EMS audit. Our emphasis should be placed on assessment of the environmental side of the use of such procedures, including environmental responsibilities, authorities, data, documents, records, and operational areas utilizing these procedures. These are also likely to be some of the key changes to the existing procedures to evaluate and determine their effective use within the EMS.

Line 7.0-Submission of EMS documentation

All relevant data, documentation, and procedures should be obtained by the audit organization with ample time to prepare, plan for, and review such documentation effecting the EMS audit. Several interesting determinations may be made by use of this specifically requested documentation prior to the EMS audit. Obtaining a listing of the environmental procedures, documents, and records controlled by the EMS is of great assistance, and so, too, becomes the review of the EMS documentation section (ISO 14001, 4.4.4) of the provided material. Understanding the core elements of the organization's management system, and their interaction and use of related documentation is one of the initial steps in developing comprehensive audit plans and protocol. The relationship of this documentation becomes a critical part of the audit process. The EMS auditor must develop a strategic plan and approach for verifying the interaction of the system and how it is effectively understood and implemented by the organization.

Summary

The most important thing to remember regarding the review of questionnaires is that they are merely examples of types of information to be obtained prior to planning and preparing for the EMS audit. Audit organizations must always consider the development of questionnaires specific to their needs and tailored to the type of EMS audit and the intended audience. The need for several different types may be necessary to cover follow-up audits, initial environmental reviews, and internal EMS audits. One last key point to keep in mind—much of this information may become frivolous and "fall by the wayside" if management of the audit organization does not provide adequate resources, personnel, and time for using such information during the planning and preparation of the EMS audit.

Development of the EMS Audit Plan

Now that we have a complete understanding of what type of information is essential, we begin the development of specific plans for conducting the EMS audit. Let us take a moment and review the guidance given in ISO 14011, section 5.2-Preparing for the Audit. The following is a specifically designed EMS audit planning form designed against ISO 14011, section 5.2.1-Audit Plan.

EMS Audit Plan
Plan No.   **Audit Date(s)**
A. Audit objectives and scope
B. Audit criteria
C. Organizational/Functional units to be audited

D. Contact person	**Department/Location**	**Responsibility**

E. High-priority elements
F. Audit procedures   **Revision:**
G. Communication of observations/findings during the audit
H. Reference Documents

I. Time/duration of audit	
J. Date	
Locations	
K. Audit team members	
L. Daily meeting schedule	
Post-audit conference (Exit interview)	
M. Confidentiality requirements	
N. Reporting requirements	
Forms (attached)	
Date of issue	**Distribution**
O. Document retention	
Audited Organization	
Audited Organization	
EMS Auditor (or prepared by)	**EMS Audit Organization**
Manager (Review and Approval)	

From information gathered from the EMS audit questionnaire for ISO 14001, the second or third-party audit programme may now proceed with the development of the EMS audit plan. In some incidences, telephone correspondence may be essential in clarifying information received in the questionnaire to further the development of the plan.

For EMS internal audits, one might question as to how practical it is to go to this extent in planning and preparing for the audit. But remember the concept of building audit protocol and files for future review and use. By developing comprehensive audit plans initially, much of the pre-audit planning is already done and available. To review, update, and modify each developed audit plan the next time around will certainly speed up the planning processes for future EMS audits.

While most of the information in the EMS audit plan is fairly self-explanatory, let us take a moment to review some of the key bits of information contained in the plan and its uses during the audit.

E.   High priority elements

Determining those elements of the auditee's EMS that are of high audit priority may be pre-requested by the auditee's organization or determined by the EMS audit programme, based on information received in questionnaires and communications regarding the audit.

In the early stages of the EMS development process, elements such as document control, objectives and targets, EM program, and monitoring and measurement may become high priority elements because of the magnitude of effort and control behind the effective implementation of these elements facility-wide. One must also consider the environmental importance of the activity. Elements concerned with continuous improvement and prevention of pollution may be targeted to strengthen the ultimate visibility to management on progress and results of such programs.

For the EMS audit programme, the results of previous audits are certainly a likely candidate for determining priority. If a significant number of EMS nonconformances are detected under a given requirement, emphasis should be placed on increasing our previous investigative and sampling processes to look into this high priority requirement more thoroughly.

High-priority elements and the information contained within each EMS audit plan become a key area for review, update, and modification for each follow-up audit to be conducted.

F.   Audit procedures

The development of audit criteria, protocol, and checklists specific to the organization's EMS and individual elements and procedures therein is essential. Chapter 5 contains pre-developed checklists specific to the requirements of ISO 14001 which are used to determine if procedures and systems meet the requirements and intent of this standard. After this pre-audit review comes the development of specific protocol/checklists in determining that the system and procedures are effectively carried out and implemented by the organization.

In reviewing and assessing procedure conformance, taking the line item statement and turning it around into a question becomes one of the most effective methods for verifying that the organization is doing what they say they are doing. In some cases, it may be detected that the procedure in question does not cover certain line item requirements of an element of ISO 14001. The same type of questioning technique also holds true here also.

"How are the handling and investigation of preventive actions carried out?" may be a question asked in the audit of such information is not disclosed on how this is accomplished within the procedure. This is, again, another illustration of a "loaded question" previously discussed in Chapter 5. The possibility exists that the organization may not be performing this task as required. However, we must verify this physically during the course of the EMS audit to ensure objective evidence has been obtained.

We reflect back again to the concept of building audit files for future use. When extensive audit procedures and protocol are initially developed for verification against the organization's EMS procedures, they may be reused over and over again. The ability to review, update, and modify our audit procedures, protocol, and checklist becomes limited after a course of time when it is determined that procedures are in place and meet the intent of the applicable requirements.

M. Confidentiality requirements

Assuring the integrity, information, and results of an EMS audit should be agreed upon between the audit organization and the auditee prior to start-up of the audit. Information deemed as confidential by the auditee should be disclosed and safeguarded in accordance with predetermined arrangements agreed upon by both parties. This becomes especially important in the reporting process. State-of-the-art methodologies, environmental engineering design concepts, and information regarding issues of potential legality and liability that may come under review and detection during the audit are examples of confidential issues worthy of consideration. The ultimate responsibility for disclosure of such information is at the discretion of the audited organization and may be based on and driven by corporate, business, and/or legal practices and policies.

Predetermined arrangements on issues of confidentiality may also be referenced in section N under reporting requirements, where applicable.

Audit Team Coordination

During the stages of planning for the EMS audit, the audit team personnel must be determined. Several key elements come into play for coordinating this team effort, including:

- determining the number of auditors needed based on the scope, purpose, and intent of the EMS audit;
- selecting of the audit team chairperson (lead auditor);
- determining lead auditor roles and responsibilities;

- determining which areas/requirements team members will audit based on background, technical knowledge (environmental or quality), and experience;
- establishing time frames for completion of audit to each area/ requirement.

The following are examples of predetermined breakdowns of a two- and three-person EMS audit team:

Two-Person EMS Audit Team

EMS auditor #1 (lead auditor)

- environmental policy
- environmental aspects
- legal and other requirements
- objectives and targets
- environmental management program(s)
- operational control
- monitoring and measurement
- nonconformance and corrective and preventive action
- records

EMS auditor #2

- structure and responsibility
- training, awareness, and competence
- communication
- EMS documentation
- document control
- emergency preparedness and response
- EMS audit
- management review

Based on information previously addressed, note that auditor #1 should certainly be the member with an abundance of environmental technical background, with sound experience in conducting audits as the chairperson. Auditor #2 may effectively audit these elements if their primary background in auditing was within that of quality systems auditing.

As we review an example of a three-person EMS audit team, note that the duties of the lead auditor will decrease. Discussions to follow will address that the role of the lead auditor will require a variety of other administrative audit functions over and above their areas/requirements to physically audit.

Three-Person EMS Audit Team

EMS auditor #1 (lead auditor)

- environmental policy
- environmental aspects
- legal and other requirements
- objectives and targets
- environmental management program(s)
- monitoring and measurement

EMS auditor #2

- document control
- training, awareness, and competence
- EMS documentation
- operational control
- emergency preparedness and response
- nonconformance and corrective and preventive action

EMS auditor #3

- communication
- structure and responsibility
- records
- EMS audit
- management review

Chapter 5 disclosed information on the use of the Universal Audit Checklist (UAC) and the concepts of exchanging and sharing data with other auditors for further review, investigation, and verification. In using this breakdown, the concept may be conducted by auditor #2 (two-person EMS audit team) and auditor #3 (three-person audit team). This will allow for more effective coverage within document control, records, EMS audit, and training within the time frame allowed for the audit.

Roles of the Lead Auditor—Planning

When serving as the audit organization's team chairperson, lead auditors must consider their roles they may be involved with outside the on-site audit process itself, including:

- selecting team members;
- providing training and guidance for inexperienced auditors during preparation and planning;
- coordinating of the team effort;
- establishing daily agendas and time lines;
- keeping the audit organization informed on progress of planning;
- assisting in the review of the audited organization's EMS policy and procedures with other team members, including the development of audit protocol/checklists;
- communicating and contacting initially with the organization regarding their EMS and the audit process; and
- conducting pre-audit site visit and tour (where applicable).

Allocation of Resources, Personnel, and Time

The author wishes to interject some observations with respect to this topic. First off, this may be somewhat "wasted" information and time spent in your study of this subject. With that, I truly apologize. However, it must be again mentioned that we rarely have enough time. Take, for example, some of the time frames that have been rather concrete for conducting ISO 9001/2 certification audits by registrars. An organization with about 500 people will probably have two auditors conducting a "comprehensive" audit of all 19/20 requirements of the standard in about two working days. That is about 28–32 audit hours. That is also about 1.5 hours spent on each requirement. Sounds about right, does it not?

For internal quality audits, registrars have found it acceptable for an organization to audit the equivalent of each requirement once every year or two, in some cases. And most of these audits last, on average, about one-half day to 1 day (including all time for planning, preparations, audit report writing, follow-up, and close-out activities).

The author wishes to make one thing abundantly clear—this is not adequate. It is just the way it is, in most cases. Could you imagine if certification audits with a facility of 500 people or so had to undergo a five-person, five-day management systems audit?

Have you ever really thought about what drives the true worldwide acceptance of these management systems standard and the certification/

registration processes? Do you think these processes would be as widely accepted as they are if the certification costs were three to five times what they are now?

So what does this all have to do with planning for the EMS audit?

For one, it is safe to say that most second-party auditors and all registrars (third-party) have significantly more time to prepare and plan for an audit than internal auditors. Remember, this may be their job—30–40 hours a week. The author truly does not intend to take anything away from the integrity and professionalism of many genuinely qualified internal auditors of management systems. To this end, the concept of extensive audit preparation and planning to initiate the EMS audit programme is critically important, (i.e., building auditor and audit program files, and creating reusable audit procedures, protocol, and checklists).

Shining through as certainly one of the core elements of the EMS is the EMS audit process itself and the organization's true commitment to continuous improvement. The EMS audit programme must always consider the allocation of resources, time, and personnel for the planning processes. It should be consistent with the depth of the investigation into the system, requirement, and/or areas being audited. A good, general rule of thumb for planning is to allow as much time for EMS auditors to plan and prepare for each audit as the time allocated to actually conduct the audit.

For example, one-half day to plan, and one-half day to audit—one total day of audit time. Now take that one day and allow a minimum of 20% of that time for report writing and follow-up/close-out operations. The later must be considered and predetermined to release the EMS auditor from "other duties" at a later time to effectively ensure the appropriate corrective and preventive actions have been taken on cited system nonconformances.

A final word for our EMS auditors—the author is aware that "it is not our call" in the allocation of time for any part of the audit process. This is in the hands of a coordinated effort between the management of the audit programme and the management of the activity/area where that auditor is coming from to support the audit programme. Somewhere, sometime, an organization should conduct an "auditor poll," taking into consideration whether or not auditors have enough time to do everything they need to do. The numbers would be staggering. Nevertheless, we strive forward to make the best out of any audit and to represent the needs of the audit programme and processes as dictated by management.

Transportation/Accommodations

The following information must be considered and planned for prior to execution of the EMS audit, where applicable:

- travel arrangements (air, car, rail)
- hotel accommodations
- per diem/expense monies
- transportation on-site/off-site
- expense/travel reporting processes

In planning all stages of the EMS audit, an in-depth personal or team itinerary should be developed, documented, and submitted to management of the audit organization. This itinerary should be developed, at a minimum, 30-60 days prior to any "off-site" EMS audit. A constant review of the itinerary from each audit should take into consideration the effort and cost behind these needs for areas of improvement, simplification, and a more cost-effective method for the supply of these needs.

Summary

This concludes "Phase 1" of the EMS audit planning process. The following chapter on scheduling the audit will coordinate and finalize the completion of these planning processes into the appropriate notifications, daily agendas, and schedules. There is still ample work to be done to effectively execute the audit and communicate these preparations to the client and audited organization. Understanding the specific sequence of these planning processes and the time involved to carry out these tasks remains the fundamentally most important element of the success and execution of the EMS audit.

With the exception of the EMS audit questionnaire, effective EMS audit planning previously discussed is equally applicable to the internal EMS audit process. The distinct advantage internal programmes have is the "reusable data" developed for the first cycle of audits against the EMS. To effectively build those audit files, the following chapter will address specific tasks to the internal processes to allow for a more cost-effective, efficient programme for conducting the internal audits consistently and effectively each time around.

8

Scheduling the Audit

Auditee Notification

Whether an internal or external audit, some form of prior notification of intent is required. An audit is a formal and usually resource intensive activity. The organization to be audited, as well as the auditing organization's management, need some notice of the purpose, scope, objectives, and schedule of the audit so that adequate support resources may be made available.

It is this author's opinion that the notification of intent to conduct an audit or, for that matter, an assessment, should be generated by the audit team chairperson, the lead auditor. It should be submitted to the audited organization as early as possible before the actual arrival date at the organization's facilities. If the audit is well-planned and properly scheduled, the notification should be received by the auditee 15–30 days prior to arrival of the audit team at the auditee's facility.

In the new world of EMS and the ISO 14000 standards, the notification will probably be addressed to the appointed management representative. It is the representative's responsibility under ISO 14001, subclause 4.4.1, to "ensure that environmental management system requirements are established, implemented and maintained" in accordance with the standard. What better person to inform of your intent to audit the EMS?

On occasions, the notification will be addressed to the top management position. This may be based on contractual requirements, established protocol, the level of EMS maturity of the organization, or other business factors. In either case, the notification must accomplish a few things.

External Audit Notification

An EMS audit notification should be similar in nature to the example form provided here. It should identify the type of audit to be conducted, such as a pre-assessment, ISO 14001 systems audit, certification audit, or surveillance audit. It should confirm the scheduled dates for the assessment/audit and state the time and date for the arrival conference. The notification should identify the key auditee personnel who should plan to attend the arrival conference. A brief outline of the arrival conference agenda should be included.

In the case of our example, the agenda consists of an introduction of auditor and auditee personnel, identification of the EMS lead auditor and his or her topics for discussion (scope and purpose of the audit, audit procedure(s), audit agenda, tentative time and date for the post-audit conference), and an opportunity for statements by the audited organization's personnel.

An EMS audit notification should describe the audit time table, including dates and the hours audit activities will be conducted. It should propose a time each day for debriefing of the auditee's management on the status of the audit as it progresses. The type of audit, such as a two-person, three-day audit or a three-person, two-day audit, should be described. And a copy of the audit's agenda should accompany the notification. These pieces of information will allow the audited organization to prepare the necessary resources for a smooth flow of the audit.

A good rule of thumb is to inform the auditee of what documents the audit team has in their possession. Our example identifies the auditee's environmental policy, level II procedures, programme documents, and a listing of the samples of environmental documentation and records.

The more information related to the EMS an audit team has in its possession to review before conducting an audit, the more effective and efficient the audit will be. But the information reviewed must be current, correct, and complete. Our notification example includes a note to the auditee to forward any later revisions of the documents to the team as soon as practicable (e.g., within 15 days of set audit date).

Always verify the correct spelling of the responsible party to whom the notification is to be sent! Few things irritate people more than misspelling their names or omitting a title (Dr., Jr., III). For that matter, check the spelling of all names before you send out the notification. And always thank the auditee for receiving and reviewing the notification. Politeness is an essential ingredient of professionalism, as we discuss later in this guide to successful EMS auditing (Chapter 10).

The notification should have a document number, any revision duly noted, and, of course, a date. In our example, we also include the review approval signature, which adds value, in the author's opinion. As designed, our example is a model of a controlled document. As stated in subclause 4.4.5, Document control, of ISO 14001, documents need to be legible, dated with dates of revision, and readily identifiable. They are to be reviewed and approved for adequacy by authorized personnel. Our notification form meets the test.

Q.S.D. Registrar, Inc.
ISO-14001
EMS Audit Notification

To: Management Representative   August 12, 1996
     DEF Corp.

Type Audit: ❏ Preassessment   ❏ ISO-14001 Systems audit
            ❏ Certification audit ❏ Surveillance audit

1. This notification shall confirm our scheduled date(s) for the EMS assessment/audit of the DEF Corp. in accordance with ISO 14001.

2. An arrival conference will be held at 8:15 A.M. on Monday, November 11, 1996, with DEF Corp. EMS management and personnel covering the following:
    A. Introduction of personnel
    B. EMS lead auditor (QSD) relates:
        1. Scope and purpose of the audit
        2. Audit procedures
        3. Agenda
        4. Post-audit conference (tentative at 9:00 A.M., Thursday, November 14, 1996)
        5. DEF Corp. statements

3. EMS audit general time schedule:
    Monday (Nov. 11) 8:45–5:00 P.M.
    Tuesday-Wednesday (Nov. 12–13) 8:15 A.M.–5:00 P.M.

4. Daily debriefing to DEF Corp. management–5:15 P.M. Monday–Wednesday.

5. QSD auditors, Michael Ross (lead auditor) and Don Sayre (auditor) will conduct the 3-day audit in accordance with the EMS audit agenda enclosed (Table C).

6. We are in possession of the following DEF Corp. documents:
    A. Environmental policy
    B. EMS procedures (Level II)
    C. Environmental management programme documents
    D. Sample listing of environmental documentation (8/12/96) and environmental records (8/12/96)

Please note: If there are later revisions of any applicable DEF Corp. documents in our possession, please forward to our offices within 15 days of set audit date(s). *Thank you.*

Internal Audit Notification

For internal EMS audits, the notification may take a slightly less sophisticated format, as suggested in our example notification form below. The content is similar but with a few differences worth discussing.

Internal audit notifications are best addressed to the person responsible for the area or activity to be audited. In most cases, internal EMS audits will be smaller in scope than external audits. Our example notification is addressed to G. Hale, the Information Management System Document Control Coordinator. (In real life, G. Hale is Greg Hale, former editor of CEEM's *International Environmental Systems Update* monthly publication) The notification is from B. Cratchet (some Dickens character, you may recall), the EMS Audit Coordinator. The subject matter is, of course, notification of EMS Audit.

The notification provides the scheduled date of the one-day audit, its scope and purpose, and applicable EMS procedures. It also identifies the areas and activities to be audited specifically, noting the scheduled hours of the audit in each area or activity. Key responsible personnel within the scope of the audit are called out as well as the notification distribution list.

In our example, we show space for acknowledgement signatures. Our fictional internal audit procedure requires a copy of the signed notification to be returned by the audited organizations. This allows for contingency planning or changes to the audit agenda and also provides a record of the notification being received.

The intent of the record is to provide some indication to the audit team that, if the signed party is unavailable for the audit as scheduled, someone will be delegated responsibility for representing the area or activity. Note the polite request for return of the acknowledged notification at the bottom of the form.

**EMS Internal
Audit Notification**

To: G. Hale   **May 13, 1996**
       IMS Document Control Coordinator
From: B. Cratchet, EMS Audit Coordinator
RE: NOTIFICATION OF EMS AUDIT
Scheduled date: June 3, 1996
**Scope/Purpose: To verify conformance with document
                   control plans and procedures per ISO 14001,
                   Section 4.4.5, Document control.**

IMS Procedures—documents

Title: "IMS Document Control," Proc. No. IMS-DC-007, Rev. C (4/15/96)

Title: "Environmental document listing," Doc. No. EMS-DC-007A, Rev. B (4/15/96)

Areas/Activities to be audited:
1. Manufacturing Department (9:00 A.M.–10:30 A.M.)
2. Heat Treat Department (10:30 A.M.–12:00 noon)
3. Raw Material Storage (1:00 P.M.–2:00 P.M.)
4. Environmental Management and Waste Minimization (EM-WM) (4:00 P.M.–5:00 P.M.)
5. Environmental Engineering (5:00 P.M.–6:00 P.M.)

Key Personnel:
All departmental managers
(Indicate names of key personnel)

Distribution of Notification	Acknowledgement
Manufacturing Manager	_____
Heat Treat Supervisor	_____
Materials Control Manager	_____
EM-WM Manager	_____
Chief of Environmental Engineering	_____

Please return notification to EMS Audit Department by 09/29/96. *Thank you.*

EMS Audit Schedule(s)

In the author's Advanced EMS auditors course, we divide an audit into three phases. Within Phase I, there are three steps:

1. Select and schedule facility
2. Select audit team
3. Contact facility and plan audit

❂Select and schedule the facility for audit❂

Selection Criteria

The selection of which facility to audit and when is based on many factors. For third-party auditors, it will most likely be preset and contracted. For second-party auditors, it may be based on what activity or product or service a facility is providing, a pre-approved audit schedule, or known problems with the facility. First-party audits are established by internal audit schedule.

The ISO 14010 guidelines for environmental auditing recommend that an audit only be undertaken if it is the lead auditor's opinion, after consultation with the client, that:

- there is sufficient and appropriate information about the subject matter of the audit,
- there are adequate resources to support the audit process, and
- there is adequate cooperation from the auditee.

These guidelines are helpful in the selection of facilities to audit and their order of precedence. No audit should be performed before its time. Remember that an audit is a planned, systematic, documented verification process of objectively obtaining and evaluating evidence to determine where specified activities, events, conditions, or management systems conform with criteria and then communicating the results of the process to responsible parties.

For new facilities or facilities new to ISO 14001, you will want to perform the audit only when they have gotten far enough along in implementing the standard to be able to demonstrate how they are doing. If the organization to be audited has only recently implemented their EMS, then it may prove more effective to postpone the audit for 3–6 months, until the organization has had time to get its environmental management feet wet. If the organization has been implementing their EMS for over a year, there should be no problem or impact on effectiveness by scheduling an audit of any of the requirements of ISO 14001.

Priorities

Priorities may be assigned by the auditor's management, an interested party, or the organization to be audited. They may also reflect known problems with the facility or organization to be audited. A priority may be a result of the audit schedule. For instance, suppose the schedule requires an annual audit of a certain requirement of ISO 14001 and the date of the new audit must, as a minimum, cover at least this one requirement, but other requirements to be audited have a later due date; then, the priority requirement would be identified. Also, the results of previous audits will usually influence setting the priorities.

✪Select the audit team members✪

Availability

Availability of key audit personnel is always a factor. The sooner you as audit team chair (lead auditor) can contact personnel for joining the team, the more likely they are to be available. Early checking on availability allows the individuals to "pencil you in," so to speak. Remember, the successful EMS audit is carefully planned and scheduled.

Once you have enlisted the personnel for the audit team, be sure to periodically reconfirm their availability and have contingency plans for the outside possibility their availability changes. People do occasionally have to take themselves out of the available arena, due to higher priorities, illness, family matters, or unplanned or unscheduled incidents.

On most EMS audits, the team will need to stay structured for the complete duration of the audit. But, sometimes one individual's expertise or the scope of the audit may only require a short duration, perhaps a day, to observe a key activity or to review and interview selected documents and personnel. If an auditor or technical specialist is only required for part of the audit, then arrangements need to be made to represent the individual's audit results at the exit interview, discussed later in Chapter 11.

Travel and Lodging

Travel and lodging arrangements are important, though seemingly trivial matters. For this reason, they are discussed later in their own subsection of this chapter.

Assignments

We discussed EMS audit planning in depth in Chapter 7. In that chapter, the audit team had obtained background information from the auditee, had defined the scope, and had determined the applicable requirements, noting priority topics and expected protocols. Earlier in this chapter, we provided a means of notification of the audit to those affected. So, now that you have scheduled the audit, have your audit team assembled, and have made the necessary travel and living arrangements, it's time to assign audit responsibilities to the team members.

The audit team lead assigns all audit responsibilities. Consider the scope of the audit and the expertise of the auditors, combining related requirements of ISO 14001 under the same auditor (like documentation and document control, document control and records, structure and responsibility, and training). Consider what you know about the auditee's audit programme, and if you have yet to obtain sufficient background information (via the pre-audit questionnaire), obtain it before finalizing responsibilities.

Now, with all this information, make sure you have the right and "essential" resources, to paraphrase ISO 14001, to perform the audit. Resources are human resources and specialized skills, technology, and financial resources.

In assigning audit responsibilities, keep in mind the guidelines of ISO 14010. To ensure objectivity, members of the audit team should be independent of the activities they audit. They should be objective, free from bias, and free from conflict of interest. They should possess an appropriate combination of knowledge, skills, and experience to carry out the assigned audit responsibilities.

Some auditors can perform well auditing any activity, particularly in management systems audits. These individuals have the necessary skills and insights into management as both an art and a science. Some auditors have limitations on their oversight abilities and work better in some areas rather than in all areas. Certain auditors may have very limited capabilities, specializing in one or two areas of expertise, although satisfying the minimum requirements for qualifying as an auditor. The lead auditor must deal with these considerations as well.

A reminder of what was said in Chapter 1: only 80–85% of trained, experienced auditors are suited for all of the demands inherent in an audit. Management (in this case, the lead auditor) must recognize the personal traits, knowledge, and attitudes required for a successful EMS audit. Auditor characteristics are discussed further in Chapter 10.

✪Contact the facility and plan the audit visit✪

Discussion

As discussed in Chapter 6, in a well-planned audit, the initial contact with the facility to be audited has been made well before scheduling. Chances are the auditee's representative was the "ISO Coordinator" or management representative. Preliminary review documentation has already been received, reviewed, and digested by the audit team or, at least, by the audit team chairperson.

In scheduling the EMS audit, the audit team chairperson needs to contact the audited facility, most likely by telephone, to make the final arrangements for the arrival, the pre-audit conference, audit, post-audit conference, and departure. Now is the time to go over the audit notification list of "to do's."

Any changes that must be made to the on-site audit agenda should be made now and relayed appropriately to audit team members affected. Let us say that, for reasons beyond the control of the auditee, the pre-audit entrance briefing needs to be slipped half an hour or, perhaps because of an unplanned outage, the auditee's key personnel will be thoroughly involved with start-up, and the audit should be shifted to a later date. A death of a beloved member of the auditee organization may affect the time of the audit. Any number of unforeseen events may occur at any point in time.

The EMS audit, even more than any other type of audit, must be flexible and the auditors, likewise. You want to conduct a successful audit, one that meets its objectives, fulfills its scope, and is accomplished as planned and on schedule. Flexibility will give you the advantage. But not to the extent of being so loose that every single unforeseen impact may be accommodated. No. The audit process is structured, like any other process, but with some leeway built in.

A checkoff of the activities and expectations outlined in the audit notification and any mandatory alterations will add assurance to conducting and completing the audit as planned. Once the contact has served its purpose, the auditee is comfortable, and the audit team chairperson is comfortable, a copy of the revised audit visit plan should be given to audit team members. Everyone needs to work from the same script at all times in an EMS audit. One score of music, one conductor, so to speak.

Priorities

During the conversation (again, communication between the auditor and auditee must be a dialogue), it is wise to discuss any priority issues to be addressed by the audit. Unforeseen, unexpected events may impact the choice, or even the applicability, of priority issues, and changes need to be made. New issues may have surfaced that outweigh existing priorities. Stay flexible, but not outrageously so. Be diplomatic, but not a wimp. Be democratic, but not political.

Protocols

Changes to the audit visit plan could require changes to identified audit protocols (interviews, reviews, escorts, audit responsibility assignments). The team leader should coordinate the changes to the plan and resulting changes to protocols. Pass the changes on to the audit team to keep them informed. There is nothing worse for an audit team member than to be surprised at a pre-audit conference that his or her individual planned assignments have been altered.

Surprise Audits

There cannot be such a thing as a surprise audit! Never! No way! No how! Do I make myself clear?

An audit is a *planned* assessment. It is not a planned surprise party! You may conduct a surprise surveillance, but never a surprise audit!

Consider reasons against an unscheduled or secretly scheduled audit: availability of responsible auditee personnel, the impact on the auditee's perception of the audit team, the impact on responsiveness and cooperation by the auditee, and the resulting impact on the quality and success of the audit. Never forget that Total Quality Management is embodied in the ISO 14000 standards. We are all in this together and should never try to trip each other up.

A surprise audit is a negative activity and will have negative results. Would you enjoy an unscheduled examination by a proctologist? A surprise visit from the IRS? A state trooper with a radar detector dressed like a tree? Think about it.

Daily Agendas/Activities

As for the daily audit agenda of activities, the author offers a simple one-page example below. The agenda identifies the audited organization, the audit team members and their areas of audit, the daily schedule, including, in this case, second shift review from 6:00 to 9:00 P.M., and the proposed schedule for the post-audit conference (exit interview), audit team travel, and report writing.

Most third-party auditors complete their report prior to leaving the audited facility. The same could work well for first-party audits but will probably not work well for second-party audits.

With regard to first-party audits, a colleague of the author's with over 20 years of audits and assessments to his credit prefers to draft the audit report on a portable computer at the auditee's facility while the audit is under way. He says he collects daily input from team members, either on computer disk or hard copy or from notes of conversations with team members. Then, by the time of the exit interview, the report has been fully drafted.

Second-party audits are usually constrained on report completion by contractual or other business concerns. Although the report may be drafted in full prior to the exit interview, chances are it will need to go through a review and clarification process back at the auditor's organization.

As you can see from the agenda on the next page, the document has a title, a document number, status of revision and its date, and bears the review approval signature of the person responsible.

The author has used a form similar to the internal audit system schedule log included here, to track a variety of audits. This format allows for scheduling

EMS Audit Agenda

Organization	Arrival conference	Daily debriefing	Post audit conference
DEF, Corp	Monday, Nov. 11, 1996 (0815 AM)	Mon-Wed (1715 PM)	Nov. 14, 1996 (0900 AM)

EMS Auditor	Monday	Section	Tuesday	Section	Wednesday	Section	Thursday
Michael Ross (lead auditor)	AM Environmental aspects	4.3.1	AM Environmental Management Program	4.3.4	AM Monitoring & Measurement	4.5.1	AM Post Audit Conference
	Legal and other requirements	4.3.2			Environmental Policy	4.2	
	PM Objectives and Targets	4.3.3	PM Operational Control ✪	4.4.6	PM Records	4.5.3	PM Travel Report Writing
					Nonconformance Corrective action Preventive action	4.5.2	
Don Sayre (auditor)	AM EMS documentation	4.4.4	AM Management Review	4.6	AM Document Control	4.4.5	AM Post Audit Conference
	Structure and responsibility		Emergency Preparedness and Response	4.4.7			
	PM Communication	4.4.3	PM Emergency Preparedness and Response ✪	4.4.7	PM EMS Audit	4.5.4	PM Travel Report Writing
					Training	4.4.2	

✪ INCLUDES SECOND SHIFT REVIEW (6:00-8:00 PM)

Environmental Management System (EMS): Internal Audit System

Auditing Selection □ ☒
Scheduled •
Audit Report (open) ◪
Audit Report (closed) ◪

ISO-14001	Proc. No.	Jan	Feb	Mar	Apr	May	June	July	Aug	Sept	Oct	Nov	Dec
Environmental Policy (4.2)	EP-1												•
Planning (4.3)	N/A												
Environmental aspects (4.3.1)	421	•						•					
Legal and other requirements (4.3.2)	422	•						•					
Objectives and targets (4.3.3)	P-423		•			•			•			•	
Environmental Management Program (4.3.4)	P-424 & P-424A		•			•			•			•	
Implementation/Operation (4.4)	N/A												
Structure/responsibility (4.4.1)	431						•						•
Training (4.4.2)	IMS 418-432			•						•			
Communication (4.4.3)	433			•						•			
EMS documentation (4.4.4)	P-434						•						•
Document control (4.4.5)	IMS 45/435		•			•			•			•	
Operational control (4.4.6)	Various			•			•			•			•
Emergency preparedness (4.4.7)	ESH 64278			•			•			•			•
Checking and corrective action (4.5)	N/A				•						•		
Monitoring and measurement (4.5.1)	441				•						•		
Nonconformance and C/A, P/A (4.5.2)	IMS 414-442				•						•		
Records (4.5.3)	IMS 416-443					•						•	
EMS Audit (4.5.4)	IMS 417-444					•						•	
Management Review (4.6)	IMS 4.1.3-45		•			•			•			•	

and tracking the audit through reporting. It is designed as a simple tool when an electronic schedule is unjustified or cost-prohibitive.

As shown by the example schedule, each of the criteria of ISO 14001 is captured along with the corresponding EMS procedure number, where applicable. Note the distribution of audits and the relationship of audit activities— for instance, in February, audits are scheduled for objectives and targets (4.3.3), the environmental management programme (4.3.4), document control (4.4.5) and management review (4.6).

As shown later in Chapter 10, there are areas of ISO 14001 that apply to virtually every other requirement of the specification. These areas are usually audited more frequently, such as document control, which is a generic process that spans most management systems. The correlation of management review to document control is based on the requirement of 4.6 to collect the necessary information for evaluation. The correlation of management review to objectives and targets and the environmental programme is from the requirements of 4.6 for the review to address the need for possible changes to objectives or other elements of the EMS.

Personal Time

Auditors require some time each day for their own personal reasons. In fact, the author believes that, barring the occasional team gathering at night for a social hour, an auditor's non-auditing time is his or her own. A time to take their EMS auditor's hat off.

Some auditors, shall we say, are couch potatoes when not on duty. Some are fun seekers. Some are isolationists. Sounds like a cross section of humanity, right? Right.

The audit team chairperson will most likely be working on the audit in one way or another every waking minute of every day of the audit through the final report (see Chapter 12 on audit reporting). But, the rest of the audit team will have a few hours to collect their thoughts, support their favorite habits, and relax from the stress and strain most audits involve and the frustrations inherent in being an auditor. With this in mind, the audit team chair should accommodate the needs of the team and let them be themselves during off hours.

Of course, accountability of team members in the morning is still the responsibility of the team leader. Each auditor is responsible for being present, but the team leader is responsible for the team as a whole. Being wise in the selection of team members is the key. Pick the dependable and reliable individual. An auditor's performance includes punctuality and meeting commitments such as schedule.

Transportation and Accommodations

Now that the team has been assembled, the audit team chairperson will see to it that any travel, lodging, or other logistical arrangements are made. If the travel is long distance, it is wise for all team members to stay at the same lodging location, if at all possible. First, it reduces chauffeuring time and, second, it allows for the team to assemble in the evenings and mornings to go over the audit schedule and audit status, share observations, discuss concerns, and participate as a team.

Keep in mind that your travel agent may be able to obtain significant lodging or travel discounts when more than two people are traveling together. The author has a recipe for travel and lodging costs during an audit. What money is saved by discounts goes into the author's vacation fund, adding value to the entire family!

In most instances, individual team members will expect that they each have their own personal accommodations in a motel or hotel. The recent upsurge in suite hotels, however, might offer satisfactory arrangements to consider. Again, this may result in cost savings. As for travel, most of us who work will travel at the lowest fares we can find.

The author has flown economy, coach, business, and first class and has noticed that you are more likely to strike up a conversation with people anywhere but in first class. The author has also found that team members will, on occasion, be willing to trade their frequent flyer vouchers for a few hundred dollars. Something to consider.

Never forget to establish how expenses will be reimbursed or, if required, be advanced. For the most part, the federal per diem guidelines are very useful in determining the typical rates of lodging, meals, and incidental expenses throughout the United States.

As for travel to and from the audited facilities, the less the number of vehicles, the better. If the auditee offers transportation to and from your lodging location, it should be considered. Accepting transportation is by no means a bribe, simply a courtesy being offered. And, as EMS moves more and more into developing countries, transportation to and from an auditee's facility may involve more four-wheel capability, even hikes through jungles or rain forests, or short single engine plane flights. Let the auditee help you out if at all possible and definitely consider any offer!

In making travel and lodging arrangements, keep in mind the availability of quality restaurants for the audit team. While most lodging establishments with restaurants are more than adequate, some are not. Eating is more than a matter of pleasure, it is a habit of necessity. A good meal is better insurance that you and your team members will function at peak performance during the audit. As for lunch, usually it will be at the facility audited or a nearby restaurant recommended by the auditee. It may be that the audit team will want to work through lunch, and a catering service may be the best choice.

It is not required, and most often not efficient, for the audit team to all have lunch at the same time. The choice is up to the audit team leader. As for breakfast and dinner, these occasions allow the team another opportunity to meet collectively, whether the audit is discussed or not. In fact, the author was a delegate to the International Auditor and Training Certification Association (IATCA) conference in late July 1996. At the dinner table the first night were delegates representing eight different countries. The topic was food, and ISO 14000 was never mentioned.

Auditee Representation

It is the auditee's responsibility to provide representation. An interested auditee will be represented by whoever has sufficient authority and resources to respond to the audit as it progresses. An auditee who could not care less will provide a representative with no authority, no resources, and no power to make things happen. An auditee with issues to hide or who is simply paranoid of being audited will provide a spy as the primary representative, someone who reports every movement, every nuance, every gesture of the auditors to management.

The confident auditee will let everyone in the organization represent the organization, referring the auditors to the right people to interview, to the right location of files for review, and to the right escorts for observing activities underway.

The author has experienced all types of auditee—the good, the bad, and the ugly, so to speak. And none of them changed the purpose or objectives of the audit. It does impact the audit when the audited organization is misrepresented or poorly represented, but this is more of an impact on how the management capabilities of the organization are reflected in the audit report.

Multiple Shift Operations Coordination

The author, as an auditor with multiple layers of experience, firmly believes in going to the off-shifts during an EMS audit. It is there you will find a different environment, different perspective, different perception, and, probably, different personalities. A friend of mine worked rotating shifts at a well-known chemicals and plastics manufacture facility in West Virginia for years. He has told me about the changes to our circadian rhythm that take place just before the sun comes up. Sometimes the changes to attitude are shared by people working in close proximity. They are neither good nor bad aspects in and of themselves, but they are worth paying attention to. When do most serious accidents happen? Most emergency situations?

It is at least worthwhile for the EMS audit to look at the early hours of the second shift, usually starting at 3:00 P.M., and the late hours of the third shift, usually ending at 7:00 A.M. With any luxury, visiting these shifts in their midstream might be something to consider, as well as work over the weekend or on holidays.

Change Control

Earlier in this chapter, you were informed of the need for flexibility in scheduling an EMS audit. This flexibility should be incorporated in your audit programme procedures, with instructions on how to control changes to schedules, agendas, protocols, priorities, and any other element of the audit process.

Change is sometimes essential, in auditing as in any other activity or service. The aim of change in the audit process is to keep the process flowing along the right river bed, keeping the audits proactive and successful, making the most of allocated resources, and keeping the process current. It is not to be avoided but accepted as a condition of doing business.

Change is a part of evolution and an effective EMS audit system must continue to evolve. As the author has told his students time and time again, the world is no longer simply black and white but various shades of gray, from black to white. Keep the audit schedule as flexible as possible to accommodate the unexpected, to have contingency plans in place that work as alternatives, and to keep moving forward.

9

The Pre-Audit Conference

The Arrival

The arrival occurs when the auditor or audit team sets foot on the audited organization's property or facility to begin the EMS audit. It is an important first step in the actual conduct of the audit, for more than obvious reasons. It provides that important "first impression" of the audited location—facility condition, presence of working personnel, alert security guards, and promptness of the auditee. The first impression is difficult to overlook.

Something that must not be overlooked upon arrival is that the audit lead needs to account for all audit team members. The team leader is something of a chaperon who is always accountable for people being present at the time and place they are required to be. For the two-person or three-person team, this presents little difficulty. But if the team is larger, then the need for this head check becomes a bit more demanding.

Introductions

One of the best civilities of mankind is to introduce himself or herself at the outset of a meeting to another human being. This works less well meeting an angry or hungry wild beast, but then, the beast did introduce itself rather curtly. Rudeness works well for some species; consider the rabid animal or the enigmatic virus or energetic bacteria. But, for us humans, an initial handshake is called for and is proper protocol. In some Asian countries, you bow in respect. Bow in the United States and you indicate something else, a gentler persuasion, so to speak.

An introduction requires dialogue—you speak to me, and I speak to you. There are the pompous who will stay silent and offer their ring finger to kiss, but not in an organization or enterprise with a true EMS. The worse thing that happens to auditors is to forget the difference between dialogue and monologue or two monologues separated by direction. When you talk to me, I listen. When I talk to you, the same is expected.

The use for interpreters is becoming more and more essential for conducting EMS audits. The U.S. Department of State and the United Nations have listings of interpreters for most every language and dialect. Of course, it goes without saying that the audit paperwork so far has been properly translated into the language and dialect of the organization to be audited, if necessary. An audit could end abruptly if you get to the facility and no one speaks the language of the audit team or vice versa. Audits require communication in the same or a mutual tongue.

Guide for the Pre-Audit Conference

The pre-audit conference is a critical element to a successful EMS audit. It is the one point in time when the audit team can finalize their schedules, coordinate their activities, negotiate alternatives, and identify to the host auditee what it is they need to complete the audit.

A pre-audit conference is little different than meeting with a new client, although this time, the audit team is the client and the auditee, the provider. What are the basic steps in either meeting a new client or meeting an auditee? Introduce yourselves, get to know each other enough to converse on the same subject, and identify needed resources and work facilities. Make sure you understand what the client wants or auditee expects. Finalize any outstanding paperwork (need for current copies of this, revised copies of that, signed copies of checks and other legal tender—just kidding!). And, set aside a time to make last-minute arrangements to support the audit as scheduled.

There is a secret to holding an effective pre-audit conference, also known as an entrance interview or entrance briefing. The audit team chairperson guides the meeting, first and foremost. The chairperson goes over the audit agenda point by point until a consensus is reached by all parties, or there is an agreement to disagree, for resolution in the future.

Here are the key points to keep in mind in conducting a pre-audit conference:

1. Introduction
2. Attendance record
3. Authorization, scope, and purpose
4. Current copy of policy and procedures
5. General audit areas
6. Workspace
7. Start, stop, and lunch times
8. Informal meetings

9. Post-audit conference
10. Current status
11. Statements/questions
12. No quick tour

✪Introduce all organizational personnel.✪

At the entrance interview (entrance meeting, entrance briefing, pre-audit conference), make sure all key organizational personnel are introduced or otherwise identified, as we discussed earlier. Highlight the audit team members and their unique responsibilities.

The introductions at a pre-audit conference should be short and sweet. The audit team chairperson introduces himself or herself. The audit team chairperson introduces each member of the audit team and their areas of focus, such as "Greg will cover document control, records, and legal and other requirements; Hailey will cover policy, organization and structure, and EMS audits; Kelly, with her 30+ years of environmental science background, will evaluate regulatory compliance with applicable legislation and other requirements; and, Mike will cover emergency preparedness and response, having been involved in the uneventful recovery of TMI 2."

✪Initiate an attendance record.✪
✪Ensure everyone present gets on the record.✪

Initiate an attendance record and make sure everyone present gives their name, job title or responsibility, phone number, location, etc. The record can be as formal as a preprinted form or as informal as a sheet of note paper. Make sure the spellings are correct.

Start the attendance list with the lead auditor, then the audit team, then those members of the audited organization. A copy of the record should be given to the auditee for information and the original retained by the audit team chairperson.

✪Explain authorization, purpose, and scope of the audit and how the auditor fits into the effort.✪

Explain how the audit and audit team members fit into the effort, your authorization to conduct the audit, the audit scope, audit purpose and objectives, and any special logistical matters. Highlight any priorities.

Identify the key objectives of the audit. Remind the auditee that yours is purely an EMS audit, not an environmental compliance audit. If you have a lawyer as an auditor, with special areas to evaluate, inform the auditee. Any matter of confidentiality should also be cleared now.

✪Request a current copy of EMS policy and procedures be available for the audit team's use and review.✪

Request that current copies of EMS documents be made available for review and use by the team. Try to get controlled copies or current copies for information only. Whatever copies you obtain, make sure you verify they are current and complete.

Usually, the audit team will have access to controlled EMS documents during the audit. However, on occasion, for any number of reasons, you may be asked to audit from uncontrolled ones, information-only copies. In either case, the copies should be marked accordingly, and their currency verified against the latest controlled document control log or program.

✪State the general audit areas each team member will review.✪

State the areas each member of the team will audit, to let the organization know who to assign as escort and who should be the auditee's subject matter expert in that area. Chances are, there will be specific activities you want to observe or locations you want to tour. With the exception of a large audit team or an auditor-in-training on the team, each auditor will go his or her separate path.

✪Request workspace where the audit team may convene.✪

Request a conference room or office or some other suitable location for the team to work in and out of. If you can get a phone or two, fine; a facsimile machine, great; E-mail, super; a copier, very important; and, if possible; computer capability, today almost mandatory.

Do not expect to be treated like royalty. The author has led many audits, especially ones at smaller organizations, where space is at a premium. Do your best to live and work with what you get. As the EMS movement floods into less-developed nations, work accommodations will change dramatically. If may be your conference room is a tent just off the edge of a jungle! You may want to address the issue of workspace during your telephone contacts with the auditee prior to your arrival.

As more and more nations adopt ISO 14000, there will be more and more EMS audits in these nations against ISO 14001. Many nations will still be developing, and their audit team work accommodations may be quite limited. There will be locations in the middle of nowhere, with little, if any, true office space. Considering that EMS applies to virtually every industry of any type on the planet, it is impossible to speculate what conference or work facilities you may encounter. Learn to live with it. Work with what you have got. Adapt to the culture and the environment at the location where you audit.

✪Find out start, stop, and lunch times.✪

Find out when start, lunch, and quitting times are. If adequate, use lunch facilities at the auditee's location. If more convenient to expedite the audit, let each team member lunch on their own.

The quality of food at any location is difficult to predict. The author has enjoyed fine cuisine at audited locations and has had to default on lunch because of its quality. But, poor food or a poorly maintained eating establishment is an indicator that the commitment to continual improvement may be weaker than stated. The author believes an effective TQM program means a pleasing food facility for the employees. Eating is, after all, an environmental aspect we each share. Why make it significant?

The quality of water in some countries is a far cry different than in others. Be prepared. The author knows of jobs in foreign nations where the food is flown in and the chef is from outside the host nation. There is preventive medicine available for most ills that befall a tourist or traveler. Be prepared. Talk with your family or company physician before you go.

The EMS audit team chairperson is responsible for the health, safety, and well-being of the audit team members. The chairperson should ascertain that each team member is prepared for whatever nation, location, or situation may be encountered during the audit. Food, water, human comforts, first-aid supplies, emergency monies, current passports and other travel papers must be provided for. The EMS auditor of this coming millennium will truly be a voyager. Some environments are far more hostile to humans than others.

Be prepared.

There may be hostility underway in countries where you perform EMS audits. This must be considered and arrangements for food, water, supplies, and sundry items made. Also, the seasons affect some countries more drastically than others. Drought, monsoons, and, at some locations, blizzards occur routinely.

BE PREPARED!

✪State times for informal daily meetings with organization personnel.✪

State the times for daily debriefings to go over the ongoing results of the audit (observations, concerns, findings, and nonconformances). These conferences will be as informal as possible, depending on the political climate. Usually, they occur some time in the last hour before the end of the work day or the end of the shift, but, some auditees prefer the morning after. In either instance, they should be as brief as practical. The lead auditor will want to brief with the audit team before briefing with the auditee.

❂Indicate tentative time for post-audit conference.❂

Indicate a tentative time for the post-audit conference (exit briefing) and request the auditee's top management attend, if at all possible. On a two-day audit, the exit will be the last order of business for the audit team on the second day. On a three-day audit, it may take place late in the morning of the third day or in the afternoon. When it is held depends on how many findings the audit team has identified.

❂Request the organization give a current status of their program.❂

Request the auditee to give you a short current status of their environmental management program. REMEMBER, keep it short, to the point, no more than five minutes! Never let the auditee digress or try to go over the entire history of the organization, back to the early 1700s! The auditor's definition of KISS, which others know as Keep It Simple, Stupid, is KEEP IT SHORT AND SWEET!

❂Request statements/questions from the organization.❂

Request statements from the audited organization or inquire whether the auditee has any questions. Answer the questions and acknowledge the statements. The statements or questions should be only about the audit, agenda, or related auditee activities. This is a chance for auditee input, and you want it to be of benefit to completing the audit without taking time from the audit.

❂Do not request/accept tour of facility.❂

Do not request or accept a quick tour of the facility unless you have made previous arrangements and have planned for it. This can be a significant time waster. In most facilities, you will tour as you audit.

To avoid the number 13 and any superstition of our readers, we have an additional item of importance. Be certain to cover the emergency signals and procedures and the safety precautions associated with the facility. Have the auditee identify the specific individual to whom the audit team is to report in case of an accident and who to account to in the unlikely event of an emergency.

The audit team should expect to, be prepared for, and offer the opportunity to the auditee to attend off-shifts during the audit. The world is different on second and third shifts. The environment is different in the night and predawn hours. Management styles are different on different shifts as are physiological and psychological systems. The author can assure you from years of experience auditing off-shifts that they often offer a perspective to evaluating the effectiveness of the EMS not observable during normal day working hours. If possible, check out the weekend shift, as well.

Watch out for the auditee who wants to take you to dinner. Usually, it is unproductive and may be an attempt to bias you in favor of the auditee. The invitation may come up during the entrance meeting. And watch for the bleeding heart who will try to pull you aside after the entrance to cry on your shoulder about the poor way he or she is being treated by upper management.

Something to keep in mind: "off the record." Usually, most auditees recognize the unwritten code of honor that accompanies the term "off the record" and will discuss things with an auditor off the record, meaning nothing goes outside of the conversation and into the audit report. This may be impractical during an environmental compliance audit, but during an environmental management audit, it may serve a useful role and route of communication.

Facility Tour Coordinations

The author, again, recommends that any tour of the facility be made during the conduct of the audit. If, however, a pre-audit tour is justified or required by the organization's health and safety practices, then it should be made in as short a time as possible, with personal safety foremost in mind.

Like escorts, the tour coordinator should be from the auditee's organization and knowledgeable of the tour route, any hazards that may be encountered, any routine or special safety precautions and appropriate personal safety protection equipment, the emergency evacuation routes, ways to communicate should an accident occur on the tour, and other general information about the facility and its operations.

Escorts

Escort personnel are far more important, in most facilities, than usually given credit. They are as important as a guide dog is to a blind person on a busy street. You and your audit team should rely on qualified escorts to make sure you get where you want to go and return safely and efficiently. There are, however, some facilities where the function of the escort is simply to make sure you do not gain access to proprietary, classified, or privileged information or activities. These escorts seldom have the training that will insure the health and safety of the audit team is protected.

Should the management individual of the auditee wish to escort you or team members to view their activities and operations, chances are fairly good that this will be quite successful. Try to make sure that other duties of the manager will not leave you or the team stranded somewhere inside the facility with no escort out.

Your escorts should understand your native language, or you theirs. Otherwise, an interpreter may be called for. Your escort must know how to take you where you need to go and return, safely.

Availability of Key Personnel

During the pre-audit conference, the key responsible personnel were identified and introduced. At that time, arrangements were scheduled for interviews and other meetings. But, the auditee's EMS organization will have a plethora of other operational and administrative duties and functions to also attend to. Things might change, including their availability as expected. It is important, in keeping the flow going, to periodically reconfirm their availability.

If the audit includes an off-shift, it is wise to have the telephone number of who in the auditee's organization should be notified if a significant condition adverse to environmental management is identified. This could be a noncompliance with environmental legislation, a physical condition warranting emergency preparedness evaluation, a safety deficiency of imminent concern to health, or other issue warranting immediate attention by the auditee. Of course, the audit team chairperson should be contacted first, if practicable, and given the opportunity to either authorize the call or place the call himself or herself.

A limitation that must be considered is the need for an interpreter at the interviews. Your budget can only allow for one, possibly two interpreters. Schedule the interviews accordingly.

Documents Change Control

Now is a fine time to find out if the documents you reviewed prior to coming to the facility have been replaced with later revisions! Unless you are evaluating a prior occurrence or activity against prior versions of documents, you must audit the organization to the current, controlled, and complete documents. An audit to obsolete documents is an obsolete enterprise to begin with!

The audit team chairperson and audit team members must verify or validate as the audit progresses that documents to be audited and to audit against are current, properly approved by authorized personnel, correct, complete, and controlled. It may be that since you last spoke with the auditee, a document has been changed or superseded or dropped completely. Now is the time to make sure. And, it may be that a document is revised or changed during the audit. The team leader and auditors need to be made aware of the change.

Remind the auditee to keep you informed of any changes to EMS documents as the audit is conducted, and keep this matter in mind for the post-audit conference (see Chapter 11).

Agenda Change Control

As you have read in Chapter 7, changes happen. The EMS audit agenda is no exception. It is as vulnerable to change as the activities being audited. Changes to the agenda should be handled in the same manner as the original agenda—authorized changes only, same review and approval cycle, notifications of changes to the same individuals and organizations, same concurrence as for the original agenda, and equivalent distribution of the revised agenda.

Changes to the audit agenda at this or any later point in the audit must be justified and completely necessary. Keep those two qualifiers in mind: *justified and necessary*. If they are, then they must be made. Which leads to an interesting thought: why not take a look at the auditee's own control system for changes to documents, products, activities, personnel, or services that cause the changes to the agenda? The samples for audit are being given to you, so to speak, by the auditee. They are as random as any and certainly within the preestablished scope and objectives of the audit. Think about it.

Auditee Statements

A statement is just that, words spoken or written, an action, an expression of an idea, opinion, or thought. During the pre-audit conference, the audit team chairperson and audit team members want the input from the auditee, want any needed clarifications, and want confirmation that the purpose, intent, and agenda of the audit are understood and concurred with.

But don't forget the KISS principle—Keep It Short and Sweet and to the point.

The audit team chairperson controls the pre-audit conference, and rightfully so. If the auditee rambles on, if the auditee rambles off in a different direction, or if the auditee commits any of the dreaded sins of a company personality as described in Chapter 6, stop them!

The author is a firm believer, as you know by now, that an auditor never has enough time to do everything that needs to be done in an audit. Don't let a lengthy pre-audit conference take even more time away. A 15–30-minute pre-audit conference is more than sufficient to accomplish its purposes. In fact, the author has held pre-audit interviews as short as 5 minutes (no changes to agenda needed, full concurrence, no questions, brief statements, if any) and never longer than 30 minutes (with, of course, those unwanted never-to-happen-again exceptions to the rule). Most entrance briefings (i.e., pre-audit conferences) average 15–20 minutes.

Golden Rule: KISS.

Daily Debriefings

The audit team chairperson should establish a time each day at the auditee's convenience to brief them on the status of the audit as it progresses. Most often, 4:00 P.M. seems to work well, with the briefing lasting certainly no more than 30 minutes (an auditor's time is the most important asset). Sometimes, the auditee may decline the offer and wait until the post-audit conference. Sometimes, the auditee may delegate the function to an administrative member.

Most times, the briefing will be with only one person representing the auditee, usually not senior management, but another responsible party. It could be that the auditee may want each responsible manager in an area audited debriefed at the end of that portion of the audit of interest to him or her. Of course, do not forget to have an interpreter if you need one.

The EMS requirements include the designation and appointment of at least one management representative for each organization. This should be your contact for daily debriefings, or someone with even higher authority. Not only are the debriefings to keep the auditee informed, but they are to effect any necessary or called-for interim actions by the auditee. In addition, debriefings are a time to reconfirm the availability of personnel the next day as well as to reconfirm the time and location for the exit critique at the end of the audit.

The audit team chairperson should hold a daily debriefing with the audit team immediately prior to the briefing with the auditee. This way, the information to be shared with the auditee will be fresh, concurred with by the audit team leader, and truly reflect the status of the audit in progress. The briefing with the team should only take 5–10 minutes; the briefing with the auditee, not much longer.

If the auditee wishes to contradict any of the day's information, listen to the reasons why and state that all information identified during the audit will be reverified prior to the post-audit conference. Inform the auditee that any additional information or clarification which can be provided for consideration should be given prior to the post-audit conference. Of course, let the member(s) of the audit team know of the situation and stress the need to verify the observation(s). Make the auditee attendees at the pre-audit conference aware of how contradictions will be handled.

A crucial part of any audit is the verification of data and observations prior to the post-audit conference and before any results of the audit are documented and reported. Things change, and the audit must accommodate for change.

Management Recognitions

Allow and even encourage the auditee to visually point out their managers and their titles, EMS roles, and EMS responsibilities. This often helps the audit team members know who to go to for clarification or interview. It helps to know where each of the responsible managers is located at the facility, again for logistical reasons.

If the auditee has an EMS management representative (or representatives), they should be at the entrance briefing and recognized by the auditee. If they have yet to appoint a representative, you will want to know why. If there are representatives, but they do not attend the entrance, you will want to know why.

Team Chairperson Coordination

The audit team chairperson should inform the auditee that the functions and activities of the audit team will be handled by the chairperson. Any requests for information from the audit team should go through the team leader. Any requests for changes, also. In fact, any request for action, change, information, or other request to any of the audit team should be coordinated by the team leader. The team leader steers the audit through to completion.

10 The EMS Audit

Philosophy and Techniques

The philosophy of EMS, as you read in Chapter 1 when beginning your journey into conformance assessment and ISO 14001 auditing, is based on 12 fundamental principles:

- An environmental policy of commitment to improved performance and prevention;
- Awareness of pertinent environmental requirements and environmental aspects;
- Clear assignment of pertinent environmental protection roles and responsibilities;
- Environmental performance planning throughout the life cycle;
- Management discipline to achieve objectives and targets;
- Appropriate and sufficient resources, including essential training;
- Preparedness for emergency and response;
- Systematic operational control for high performance;
- Environmental performance evaluation for improvement;
- Review and audit for improvement opportunities;
- Communications, internally and externally; and
- Motivation of contractors and suppliers for improved environmental performance.

These dozen principles are the foundation of effective environmental management and, therefore, the pillars of the EMS to be audited by looking at the parts and components of the system as designed, implemented, checked, and changed.

In anticipation of the following chapter on EMS auditing and designing questions to ask and audit from, let me share some simple characteristics of the EMS that you can always question. Later, you will find some "standard" assessment checklist questions that may always be asked, no matter what requirement of ISO 14001 you are examining. But, for now, here are some key questions directed at overall conformity assessment of an organization's EMS, which may be "over and above" the requirements of ISO 14001.

Let us go through this one principle at a time, to keep it simple:

- **An environmental policy of commitment to improved performance and prevention**

  Question: Is there a documented policy, with a stated commitment to improved environmental performance? To the prevention of pollution?

- **Awareness of pertinent environmental requirements and environmental aspects**

  Question: Is there demonstrated evidence the organization has identified and has access to legal and other requirements directly applicable to the EMS? Is there demonstrated evidence the organization has identified the environmental aspects of its products, processes, or services?

- **Clear assignment of environmental protection roles and responsibilities**

  Question: Is there objective evidence that responsibilities and authority for environmental performance and protection have been clearly assigned? To competent individuals?

- **Environmental performance planning throughout the life cycle**

  Question: Does the auditee have objective evidence to demonstrate that environmental performance planning is a consideration of their process for life-cycle assessments?

- **Management discipline to achieve objectives and targets**

  Question: Is there objective evidence of a management discipline for meeting objectives by attaining environmental performance targets?

- **Appropriate and sufficient resources, including essential training**

  Question: Is there demonstrated evidence that top management provides the essential resources for implementation and control of the EMS? Is training of responsible personnel one of the allocated resources?

- **Preparedness for emergency and response**

  Question: Is the organization prepared for emergencies? The potential for emergencies? Accidents? The potential for accidents? Potential environmental impacts from either emergencies or accidents? Is the organization capable of responding and mitigating emergencies or accidents that may have environmental impacts?

- **Systematic operational control for high performance**

  Question: Are there operational controls to achieve improved environmental performance? Is there objective evidence of the controls? Could the absence of any of the operational controls result in deviation from the organization's environmental policy? Objectives? Targets?

- **Environmental performance evaluation for improvement**

  Question: Does the organization conduct environmental performance evaluation to measure and monitor performance? Is there objective evidence to demonstrate?

- **Review and audit for improvement opportunities**

  Question: Is there a programme for periodic audit of the EMS? For periodic review of the EMS by management? Is there objective evidence that improved environmental performance is considered in the audits and reviews?

- **Communications, internally and externally**

  Question: Does the organization demonstrate a process for internal communication of the EMS programme, procedures, processes, practices, roles, responsibilities, and authority? Does the organization demonstrate through objective evidence a process for receiving, documenting, and responding to relevant external communication on its EMS?

- **Motivation of contractors and suppliers for improved environmental performance**

  Question: Is there objective evidence to demonstrate the organization recommends/encourages its contractors/suppliers to implement an EMS?

These questions will never fill the entire EMS audit checklist, but they are provided to make you think. What is it you are after in an EMS audit? What is there to audit and to audit against?

Phase I—Audit Venues

Phase I of an EMS audit covers at least seven separate venues:

1. Determination of need for procedures/instructions/changes

2. Determination of extent of conformance

3. Determination of extent of conformance to contract requirements

4. Assessment of familiarity

5. Reinspection/retesting

6. Availability, use, and adherence to criteria

7. Sampling, data collection, and analysis

The typical purpose and objectives of an EMS audit are to determine whether the EMS conforms to planned arrangements for environmental management and whether the EMS has been properly implemented and maintained. Within an ISO 14000 EMS audit, an additional objective should be to determine whether there is continual improvement in environmental management, resulting in continual improvement in environmental performance.

✪Determination of the need for new procedures or changes to existing documents✪

If you find need for new or revised procedures, why did the auditee not determine such a need? In fact, if the organization has an EMS and is conducting their own internal EMS audits effectively, your audit team should find limited new problems. The need for new procedures or changes to documents such as procedures may be an indicator of a weakness in the auditee's EMS audit programme, audit procedure(s), management review, document control, or even their commitment to continual improvement.

✪Determination of the extent of conformance to established policies and procedures✪

A "conformance to policy" nonconformance should never be found by an EMS audit if the management systems are intact and working. Policy is senior management's commitment. It embraces the vision and mission of the organization and the reason they exist. A failure to conform to policy is a serious finding and deserves immediate attention and correction, with the potential for future nonconformance precluded by proper preventive action.

A failure to conform to the environmental policy may have serious legal, social, business, or other consequences for an organization. The policy is the one document required to be made public by ISO 14001. As such, it is a binding document. It is a written statement of a commitment to comply with relevant environmental legislation and regulations, other requirements to which the organization subscribes, continual improvement, and the prevention of pollution.

Procedure nonconformances are more typically found. Of course, the question of an independent EMS audit team should be, "Why were they not found and identified and corrected by the audited organization before now?" Usually, if the EMS is implemented effectively, the nonconformances identified by an EMS audit should be minor in their nature and impact on environmental performance. There may, however, be a string of these nonconformances that point to a more systematic problem. Two wrongs do not make a right, contrary to some philosophies, but three wrongs always point to a mutual source or root cause.

✪Determination of the extent to which established policies and procedures conform to contractual requirements✪

When the EMS audit covers conformance to contract requirements, the audit team should review the process and practices of the audited organization for contract review. They should look at how the contract requirements are specified and whether strict conformance is up to interpretation. If the initial contract review was carried out correctly, the organization's procedures should conform, or it may be that there is a document control problem to investigate.

✪Assessment of the familiarity of operational personnel with documentation with which they are required to perform✪

A multitude of operational documents and procedures are usually found in facilities of any complexity. These may include technical specifications, standard operating practices, instructions, drawings, vendor manuals, limiting conditions for operation documents, and others. Most operations with the potential of resulting in a significant environmental aspect that results in a significant environmental impact require familiarity with this array of documents. The various functions and levels of operations—from construction to manufacturing to maintenance to quality, engineering, storage, etc.—should be able to demonstrate sufficient familiarity to satisfy an EMS auditor.

✪Reinspection/retesting of collective samples✪

If any previous EMS audits of the organization have been conducted, your audit plan may include as priorities a reconsideration of the prior samples. The author uses the words "reinspection" and "retesting" to make a point; an audit may involve inspection and testing as well as reviews, interviews, evaluations, and examinations. It is an analysis and may involve many component methods of performing a comprehensive deduction of the "as-found" conditions of EMS implementation.

✪Assurance of availability, use, and adherence to work, inspection, and monitoring criteria✪

The EMS specification standard, ISO 14001, includes document control requirements that call for EMS documents to be located and available at all locations where operations "essential to the effective functioning of the system are performed." Current versions of documents are to be utilized in implementing the EMS. Documented procedures are required to "monitor and measure" the key characteristics of the organization's operations and activities that "can have a significant impact on the environment."

Although ISO 14001 has specified no requirement to adhere to any set criteria, it does specify that nonconformance to EMS requirements be handled and investigated and that action be taken to mitigate impacts. The nonconformance is to be corrected and recurrence should be prevented.

When an EMS audit by a second or third party identifies a significant number of nonconformances to document control requirements, the auditee's nonconformance control process should be reviewed.

✪Sampling, data collection, and analysis✪

The author's advanced EMS auditor course offers an overview of a typical 2-person, 3-day audit in which each auditor randomly samples 15 procedures and 15 records, then randomly samples the training evidence for 20 individuals. In total, this equates to 30 randomly sampled procedures, 30 randomly selected records, and 40 randomly chosen training files or records. This is a sample size for a 3-day audit that is sufficient for determining whether the procedures, records, and competency of personnel conform to requirements.

Rules to keep in mind: (1) verify that the procedures sampled are not obsolete; (2) never ask the auditee to bring you the records; and (3) choose individuals at relevant functions and levels whose work could result in a significant environmental aspect.

Why do you think you should never ask the auditee to bring you the records? Think about it. What records do you expect to get? Right, the ones that are "picture perfect." The ones from the "file for auditors" that some auditees have been known to keep just for your visit!

Phase II—Steps

In the author's advanced EMS auditor course for environmental and quality professionals, Phase II of the audit is broken down into five key steps:

1. Understand internal controls
2. Assess internal controls
3. Gather audit evidence
4. Evaluate audit evidence
5. Report findings to facility

✪Step 1: Understand Internal Controls✪

- Review background information
- Attend opening meeting
- Attend orientation tour
- Review audit plan
- Confirm understanding of internal controls

Earlier, in Chapter 7 on audit planning, we covered the review of background information. In Chapter 9, we reviewed the needs for the opening meeting and the pre-audit conference and stressed avoiding any orientation tour of a facility unless absolutely necessary for safety or to expedite the audit. Review of the audit plan was discussed in several earlier chapters, including review of the plan with the auditee for concurrence.

The confirmation of understanding an auditee's internal controls is a twofold activity: first, in Step 1, the activity is the audit team's initial review of auditee programmatic documents (first level policy, second level system procedures, third level work procedures, and fourth level records); and, second, the audit itself (observations, interviews, reviews, and samples).

✪Step 2: Assess Internal Controls✪

- Identify strengths and weaknesses of internal controls
- Adapt audit plan and resource allocation
- Define testing and verification strategies

Step 2 is one part within the audit preparation stages. The preliminary evaluation of auditee EMS documents may identify strengths or weaknesses within their internal controls. From the review, the audit plan may need to be adapted for new priorities, depth or scope, size or duration, and different resource allocation.

With these dimensions understood, you will define your audit strategies (sample sizes, material tests for environmental aspects, procedure tests for response capability, and validation in lieu of verification).

✪Step 3: Gather Audit Evidence✪

- Apply testing and verification strategies
- Collect data
- Ensure protocol steps are completed
- Review to determine that all findings are factual
- Conduct further testing if required

The actual facility or organization audit is what the author terms Step 3. Now is when you implement your strategies, collect data for evaluation, and complete your predetermined protocols. Once the auditor or audit team has reached this point, it is time to reverify what you have found and determine whether your findings are actual or suppositions. Further examination/testing/sampling may be in order.

An audit is only as good as it is planned and performed. What is observed during an audit may lead to conclusions, and these conclusions must be factual, supportable, verifiable, and reproducible. An audit is a test, in a way, and the protocols and strategies followed need to be crystal clear so that any other auditor of equal competence could follow them and come to the same conclusions.

✪Step 4: Evaluate Audit Evidence✪

- Develop complete list of findings
- Assemble working papers and documents
- Integrate and summarize findings
- Prepare report for closing meeting

On day three of a 3-day audit or day two of a 2-day audit, the audit team chairperson, with input from all audit team members, documents a listing of the audit findings. Findings are the nonconformances and concerns identified during the audit's snapshot of activities and practices in time. Later in this chapter you will learn how to identify trends in findings so that like nonconformances may be grouped where appropriate for corrective or preventive action.

Strengths and weaknesses are also audit findings and should be identified. An EMS audit is a proactive undertaking used to verify conformance to requirements and identify both positive and negative aspects of management.

❀Step 5: Report Findings to Facility❀

- Present findings at closing meeting
- Discuss findings with plant personnel

Chapter 11 covers how to hold the post-audit conference. This is the formal presentation of the audit findings to the auditee for consideration. It is the formal opportunity in the audit to discuss the findings in a forum of responsible auditee managers and other personnel.

A fellow auditor friend of the author's has an odd sense of humor and thinks this way. When you present findings to an auditee for consideration, the auditee will also need to digest them. Digestion can be done one of two ways in the biological world: aerobically or anaerobically.

In the first mode of digestion, the end results are harmless—carbon dioxide and water. In the second, methane gas results. His point is—know your audience (the auditee) as best you can before presenting findings, especially ones of adverse conditions within their EMS. Which of the two end results do you want to walk away with?

Communications

An EMS audit is one of communication. The auditors ask, and the auditee answers. When the auditee asks something of the auditor, the auditor should only answer if he or she knows the answer. If in doubt, defer to the lead auditor. There will be normal communication that occurs whenever one person is with another person.

The ISO 14001 standard requires certain information to be communicated specifically. The environmental policy is to be communicated to all employees and available to the public. EMS roles, responsibilities, and authorities are to be communicated to facilitate effective environmental management. EMS performance is to be reported by the management representative(s) to top management for review. Employees are to be made aware of the importance of conformance to policy, procedures, and EMS requirements. They are to be made aware of any significant actual or potential impact of their work and the environmental benefits of improved personal performance.

Employees are to be made aware of their roles and responsibilities in achieving conformance to policy, procedures, EMS requirements, and the requirements for emergency preparedness and response. They are, as required by ISO 14001, to be made aware of the potential consequences of departure from specified operating procedures.

Subclause 4.4.3 of ISO 14001 is on communication. It requires internal communication procedures and procedures for receiving, documenting, and responding to relevant external communications from interested parties.

During the audit, anyone with an EMS role or responsibility at any function or level of the audited organization should at least be aware of the EMS and its requirements. Not that each individual will know everything about the EMS, but each should be able to communicate on their involvement in environmental management. Most every worker in the world knows who they report to. It might be of interest to see if they know who they report to in the EMS structure.

To communicate means to express yourself and to be understood. We will show you audit questioning techniques later in this chapter to make sure you are understood and that you understand the responses. And you must be heard; when communicating in a loud facility, make certain your message has been received as you said it and as you meant it.

Remember to have access to and use an interpreter when appropriate. Not all interpreters are as efficient in communicating certain types of information as others. An interpreter good in social discussions may have less familiarity with technical terms, such as design engineering, mechanical systems, or even the environmental management systems and the wording of the ISO 14000 standards.

Sampling

The sample size an EMS audit team selects is no firm rule of thumb. You do not need the strict sizes called for by documents such as Mil-STD-105. This is not an inspection of product, but an audit of management performance.

In most audits (the 2-day or 3-day), a sample of more than 12 is usually sufficient to find inconsistencies in records, problems in procedures, or omissions in training. For facilities new to the EMS, the available data may allow for only smaller samples, perhaps as small as one. The audit will adapt to what resources are available.

Keep in mind that you are not in this for the numbers game—1 out of 5 records were found to..., 6 out of 8 or 75% of the procedures reviewed were found to..., 10 out of the 10 training files researched were found to.... No, you are here to verify conformance to requirements and areas of nonconformance. It has nothing to do with numbers. There are no quotas in EMS auditing.

Team Chairperson's Responsibilities

The audit team chairperson, as lead auditor, has numerous responsibilities. In this section, however, we will only relate some of those pertinent to our discussion of EMS auditing in light of what you have read so far:

- Team member selection
- Training and guidance
- Coordination of team efforts
- Daily agenda and time lines
- Keeping auditee informed
- Assistance in reviews
- Initial communication and continued contact
- Pre-audit visit/tour

The lead auditor is vital in audit planning as planner, scheduler, resource allocator, communicator, reviewer, interviewer, catalyst, coach, and coordinator.

Team Selection

Selection of the right types and personalities of audit team members is essential for the success of an audit. Mixing the wrong expertise or incompatible attitudes causes great strain on the audit. In an EMS audit, there are areas to be evaluated that should be handled by someone with management savvy, and others that should be handled by someone with an environmental background.

Select every member of the team. It is possible that your management or even the auditee may suggest certain individuals for the team or suggest not using certain individuals, for whatever reason. But the lead auditor makes the call.

Never take an inexperienced auditor into the field with assigned audit responsibilities without some training or personal guidance. It is like taking a nonswimmer into the ocean without a life jacket.

Training and Guidance

Earlier, in Chapter 8 and this chapter, you learned the keys to selecting individuals for audit teams and assigning audit responsibilities. But there will come a time (mark my words) when you have to train someone during the performance of an audit. On-the-job training is an essential part of auditor training. As in many professions, there is an internship period to qualify a person to fly on their own, so to speak. Intern, "rookie," apprentice, interim, associate—they are all a phase of the learning process leading from amateur to "The Professional."

Training an auditor during an EMS audit involves the apprentice participating as a helper to the lead auditor. During the audit, the apprentice will perform under the lead auditor's direct guidance and in the presence of the lead auditor. The apprentice will follow the audit checklist he or she has helped prepare or has been indoctrinated on. The apprentice will be critiqued throughout their performance and kept informed of the quality of their performance.

Coordination of Team Efforts

In most cases, it is wise to have a lead auditor with management and environmental background and experience. The lead auditor needs to make sure all team members are properly prepared through training or guidance on audits. If your audit requires technical specialists, consider them team members, but not auditors. They are technical resources to be utilized by the auditors.

Audit team leaders need to be master coordinators, like a symphony or choir conductor. At the same time, they are coaches and mentors, identifying strategies, keeping the audit team focused, motivating, not manipulating, and managing, not mandating.

Daily Agenda and Time Lines

The daily agenda is like a plan of attack. An effective audit, covering two, three, or five days, has a tight schedule for interviews, observation of activities in progress, meetings, and so on. It is important for the audit team to stick to the daily agenda, whether changes are required during the course of the audit or not.

Keeping Auditee Informed

Daily debriefings at the end of the day or first thing in the morning are effective ways of keeping the auditee posted on audit progress. Try to work around the auditee's daily schedule. Sometimes, the auditee will decline the offer for daily debriefings, but usually they want to know the status of the audit every day. If significant issues are identified, it is best to bring them to the table as soon as possible.

If the potential of a regulatory noncompliance is found during an EMS audit, we will show you how to handle it later in the book (Environmental Communications Memo, Chapter 12).

The team leader will want to meet with the audit team members just prior to briefing the auditee so that the information is as current and complete as possible.

Assistance in Reviews

It is usually best for all audit team members to know the organization's environmental policy. The purpose of an EMS is to implement the policy and commitments to meet the objectives and targets and for continual improvement. Audit protocol is determined, in part, by the purpose and scope of the audit, but there are additional protocols, such as interview techniques, sampling, and correlating facts and observations. The lead auditor is involved in these determinations and assignments and always approves the audit checklists.

Initial Communications

The lead auditor will be the first to communicate on the audit with the auditee. We went over a type of audit planning checklist earlier.

Site Tours

The opportunity for a pre-site visit is rare but may be beneficial, especially for large, complex, or multiple-site audits. If the facility is small and simple, it would be a waste of time for a pre-site visit. The presence of any imminent hazards should be considered before ruling out a pre-site visit. You want to leave a facility with your audit team intact!

Lead the Team

Lead the team effort. The management guru, Peter Block, suggests that leadership means to create order, consistency, control, and predictability. Leaders are successful when they distribute ownership and responsibility, when they balance the power and promote a partnership.

Lead the Arrival Conference

Lead the discussion in the arrival conference (the entrance meeting or briefing). You know the scope and purpose of the audit, the schedule, and the auditors assigned to specific tasks.

Represent the Team

Represent your team at all meetings with the auditee's management.

Make Final Decision

Make the final decision on what will be reported at the closing critique (exit briefing).

Stay Advised

Keep actively advised of the findings, conclusions, and recommendations for the exit interview.

Supervise and Coordinate

Supervise and coordinate. You are the "Chairman of the Board" for the EMS audit team. Oversee preparation of the individual reports of findings and the final audit report that results from your investigation, your audit.

During the kickoff interview, make a second verification of the start and stop times of the auditee's operations, and the preferred lunch hours and locations. Remember to get direction from the auditee on emergency signals, emergency exits, personal protective equipment (PPE), and the person you should to account to during an accident or emergency.

Leader and Team Player

Be a leader and also a team player. Leaders, according to Stephen Covey, the TQM guru for the U.S. Department of Energy:

- are proactive and responsible, know that behavior is a function of our decisions, not our conditions, and are able to choose a response
- begin with the end in mind
- put first things first
- think win/win
- seek first to understand, then to be understood
- synergize

Some recent management consultants have suggested that leadership is doing the right things, while management is doing things right.

A leader creates a clear, mutual understanding of what needs to be accomplished, focusing on results, not on methods.

Take Charge

An EMS audit team leader takes charge and, believe it or not, leads! You will not be out in front of the team but will lead as a part of the team. You are in control of the audit at all times throughout the audit to completion when you have accomplished the audit plan.

Decision Maker

The team leader is a decision maker, making the final call like a referee. The lead auditor is the arbitrator, choosing one decision over another. Your decisions should coordinate the findings, should reflect what the findings indicate, and should be simple statements of fact without personal opinion. There are occasions when the opinion of the audit team will need to be expressed, but always in light of the professional opinion of the audit team.

Firm but Fair

The team leader is always firm but fair, without bias, with no unsubstantiated opinion. Please disregard the pun, but, firm and fair requires flair. It is being diplomatic and balanced. It is being tenacious and direct.

Extraordinary Knowledge

The team leader must be knowledgeable of the specifications of ISO 14001. As you have read earlier in this book, the EMS audit team chairperson must also be very knowledgeable of environmental performance requirements, from enabling legislation down to standard practices and techniques. And for successful EMS audits, the team leader must know management, both the requirements and the processes.

Express Self

The team leader will express himself or herself effectively, exhibit a professional attitude and manner, and be a good communicator, both orally and on paper. This expression is not of the individual but of the individual in the capacity and position of EMS audit team leader. It is never a personal statement, but one based on the leader's knowledge of EMS, environmental performance, and management requirements.

Review and Determine

The team leader is to review the work of the audit team and determine its acceptability as the final judge of quality. Each auditor may have performed fairly independently of the other auditors within the team effort. It is the team leader who coordinates these various viewpoints into a consolidated decision for presentation, reporting, and where necessary, corrective and preventive actions.

Write Effectively

Lastly, the team leader of an EMS audit team must write effectively or know to whom to delegate this responsibility. The audit reports are to be written to document the facts as found and, where nonconformances are identified, to result in proper and appropriate action by the auditee, whether corrective or preventive.

The freedom of EMS audit reports from disclosure outside the auditor and auditee organizations has yet to be guaranteed. Confidentiality may be viewed differently by one regulator than another or one lawyer than another.

224 THE ISO 14000 EMS AUDIT HANDBOOK

A well-written, clear, and concise audit report is expected, whether it remains proprietary to the auditor and auditee, or not.

One of our fellow associates has these tidbits to offer to auditors in being both effective and efficient. First, have an interest in researching situations to find out what was done, who did it, and why—be inquisitive. Second, the talents of an auditor depend on experience in auditing and being audited (auditors make the best auditees, but not necessarily vice versa). Be curious, cordial, comprehending, competent, and courteous.

Associate EMS Auditors

There are many types of associate auditors. There is the audit team member, the auditor-in-training, the legal counsel, technical specialist, and the observer.

A full audit team member is one with sufficient competence to perform an EMS audit under a lead auditor. They may be qualified lead auditors themselves but in this capacity, they perform under the authorized lead.

An auditor-in-training is someone whose competence requires hands-on auditing experience. They will perform under the supervision of the lead auditor.

Legal counsel may be a member of the team in a purely advisory capacity, or may perform as an auditor or auditor-in-training, depending on their recognized level of competence.

Some EMS audits may involve having a technical specialist or two with specific expertise. They are members of the audit team as specialists, with no real auditing functions or need for auditing competence. They will perform under the supervision of the lead auditor.

The observer is rare to EMS audits, but certain federal agencies offer the opportunity to have experts or other regulators attend audits simply to observe. The observer has no auditing authority or function. Their competence as an EMS auditor is irrelevant in their capacity. They perform in association with the audit team but under the supervision of the lead auditor.

The interpreter is an associate to the audit team, not a member, unless one of the auditors serves as interpreter. In fact, there may be occasions or situations where a member of your audit team should be from the host nation, aware of the language, dialect, culture, and protocol.

Words to Challenge

- Tightly
- Thin/thick
- Desirably
- As applicable

- Approximately
- To ensure good workmanship
- Snugly
- As necessary
- Sufficiently
- Properly
- Good commercial practice
- Etcetera
- Suitably
- Firmly
- As required
- Securely
- Equivalent

Notice what all of these words have in common, what they share. Ambiguity. None of them lends itself to quantification or measurement. As Dr. W. Edwards Deming loved to tell management around the world, if it can't be measured, it can't be managed. And, with perhaps the exception of a romantic poem, if it cannot be grasped in terms of physical dimensions, it cannot be verified or reproduced.

Look at the word "tightly." What does it mean? How tight is tightly? The author's experience in the nuclear industry during construction of the mighty power plants of earlier years ran into "snug tight." Can that be inspected by anyone other than the person who applied the "snugness"?

"Thin or thick." Again, what are the actual dimensions? When is thin too thick or thick too thin?

"Desirably" merits a place as the closing of a letter to a lover. It has no place in an EMS!

"As applicable" leaves the determination of applicability up to… who? O.K. for novels or even some textbooks, but would you want to take the word to court? Yes, your Honor, the defendant is guilty to the charges as applicable to the exceptions cited in the defense statements. Come on, give me a break!

"Approximately." Curious word. Take the lunar lander to approximately where the moon is right now and wait for further instructions for landing.

"To ensure good workmanship" is a classic. It worked back in earlier centuries, perhaps, but we are entering the new millennium! Use the right tool to ensure good workmanship. To ensure good workmanship, ask yourself: Can it be quantified? Controlled? Understood? Reproduced?

"Snugly" is as snugly does. Make sure you are snugly in the saddle before you jump the fence. Make sure your seat belt fits snugly on take off and more snugly on landing. The lugs should be snugly tightened so the wheel does not come off at 70 m.p.h.

"As necessary" is one of the most abused terms ever. What is "as necessary"? When is "as necessary"? How is "as necessary"? Who is "as necessary"? Define it so I can get my arms around it. Run the product through the machine as many times as necessary to get the desired look. Turn the nut counterclockwise at least twice or as necessary to get it snug tight. Add as much heat as necessary to attain the desired temperature. Have as many people as necessary to control the process from achieving critical abnormal operating characteristics.

"Sufficiently." Another romantic letter closer. Sufficiently Yours. The door was sufficiently closed to limit the decibels of noise below standards. Add enough thinner to the paint to sufficiently reach the desired thickness. We have sufficiently reached a consensus not to hire you.

"Properly" smacks of pepper words, like "uppity," "perceptively," "implyingly," and "supposedly." They feel good on the tongue, but what do they tell the worker? And what do they tell the EMS auditor? It is a gray world now, yes, no doubt about it, so our terms and definitions must be clear, precise, unquestionable.

"Good commercial practice" in one industry is different in another. Same for different countries, states, cities, and towns. The federal government in the United States is as guilty as anyone of misuse of this phrase. Consider CERCLA and its innocent landowner defense—"all appropriate inquiry into the previous ownership and uses of the property consistent with good commercial or customary practice."

"Etcetera," etcetera, etcetera. Great word to dictate to a secretary for a memo: To Whom It May Concern, etcetera. Regarding your recent letter of complaint, etcetera. We at We Seek 'Em detective agency must remind you of our policy regarding refunds, etcetera. If a step in a procedure applies to more than 1 or 2 specific things or activities, spell them out. The use of "etc." is the worst shortcut in the history of management! It will lead the auditee into trouble every time. It has many interpretations for many people. It is a word that should always be challenged by an EMS auditor. The EMS has a defined scope and application that can be measured. It has boundaries, whether physical or otherwise.

"Suitably," "Firmly," "As required," "Securely," "Equivalent." Can you define what any of these words mean in an EMS document? Would you expect to see any of these words in a management review report? In an EMS procedure? In an environmental performance evaluation?

Other words to challenge—may, approved equal, all (this one *always* leads to failure), ensure, and assure.

There is a story the author heard from a nuclear industry auditor. It seems one of the projects audited ran into the term "calibrated arm." Apparently, there was a torque requirement for certain hold-down bolts on the reactor base, but the location of the bolt prevented using a calibrated torque wrench. So, the contractor qualified the arm of the millwright on a torque calibration machine. Then the craftsman fit his wrench to the bolt and torqued it until he felt the required pressure had been applied.

In summary, do these words to challenge provide sufficient guidance and instruction to allow the worker, operator, supervisor, or management to do everything they need to in satisfying a given requirement? The EMS auditor should inquire as to a consistent interpretation and challenge the organization to demonstrate "as necessary, applicable or as required."

Questioning Techniques

- Straightforward professional
- Perplexed attitude
- I don't understand
- If this happens...
- Appear dumb, not smart
- Saying nothing
- I see it says...however
- Kill them with kindness
- Unavailable data

☺Straightforward☺

This is the professional approach to questioning during an EMS or any other audit, for that matter. It means you treat others as professionals, and they, in turn, treat you as the professional you are. It involves demeanor, attitude, finesse, and the ability to communicate through questioning.

☹Perplexed☹

The perplexed approach I like to call the "Columbo technique." Those of you who have seen Peter Falk in his famous Inspector Columbo role on television can relate to this attitude. It is very useful in some situations. His key to successfully solving crimes came from his ability to always seem perplexed by what you said compared to what the evidence showed. His brow would furrow, his hand would go to his forehead as if in deep thought, then he would gesture and ask you to clarify what you had just said. "Now, correct me if I'm wrong," Columbo would say, "but...." Or he would say, "Now, let me see if I've got this straight...."

This is an important technique for the EMS auditor in conducting a second verification of previously discovered observations on findings. Utilizing this technique may assure the auditors have all the facts and data correct and understood before leaving the audited process or activity.

☹I Don't Understand☹

Be honest in your approaches as an EMS Auditor. Grey areas may come up during the audit which go beyond the technical knowledge base of the auditor. Remember, auditing is indeed a learning experience, and this is certainly an opportune time to increase that knowledge base by compiling detailed notes for future reference.

☺If...Then☺

What if this happens? This is a very important questioning technique in determining how groups or individuals may react to a situation that creates difference. One of the best times to use this technique is when an auditor has verified an individual as the *only* one at at the facility who is responsible for performing a given task. The auditor may inquire, "How is this performed if you are not here?" The EMS auditor may use this technique in many situations to determine how well the system functions if "things happen."

☹Show Me☹

Ah, the proverbial Missouri tactic.... Remember, through planning, preparation, and the development of audit protocol against the organization's system, the auditor has many of the "answers" already. An important part of our verifications is the demonstration of sufficient fulfillment of the organization's policy, procedures, and programmes. Rely on facts as a basis for your conclusions.

☺Silence Is Golden☺

If the auditee becomes agitated, frustrated, or "clams up" during the audit, there is no sense in adding fuel to the fire. The EMS auditor may elect to point out the strengths in the system at that point of the audit or move on to another area of inquiry to get the audit and auditee back on track.

☺I see it Says...However☺

This previously discussed technique may allow the auditor to question procedures which are not exactly being followed as designed. Remember to take detailed notes on the differences observed to establish concrete evidence of the deviations presented against a given set of requirements.

☹Unavailable Data☹

When data is not available at the time of the audit, report it that way. Later, if the auditee provides the missing data, fine, but at the time of the audit, it was unavailable.

In an EMS audit, the author recommends you spend no more than four hours on each requirement being audited. This is a handy time frame to have in mind in scheduling your audit. Say you have an audit of 8 of the 17 criteria in ISO 14001 planned at a facility. You would then schedule a 2-person, 4-day audit.

Documents Review—Availability

In subclause 4.4.5 of ISO 14001, Document control, the auditee's procedures are to control all documents required by the standard so they can be located and are available at all locations where essential EMS operations are performed. The documents are to be reviewed periodically, revised when necessary, and approved for adequacy by authorized personnel. Documents are to be legible (can be read), dated, and readily identifiable (what the document is for).

What documents are required by ISO 14001? Well, any document the EMS organization says is required, for one. But, as a minimum, the standard calls for a documented policy, documented objectives and targets, and documented roles, responsibilities, and authorities. It requires documenting communication from interested parties relevant to environmental aspects and the EMS. It requires certain documented procedures and documented management reviews.

There is a subclause of ISO 14001 that is titled "EMS documentation, 4.4.4." This section of the standard requires paper or electronic media information that describes the core elements of the system and their interaction. This documented information is also intended to provide direction to related documentation.

Document control, as you will see later in this chapter, is one area that may be an integrated system. Some organizations will have one document control system for all types of documents. An audit of document control will usually find at least some minor problems.

Data Collection and Review

The prime functions of an EMS audit are the collection and review of data—objective evidence. Objective evidence includes the lack or absence of data, also. Data collection should be by random sample, as discussed earlier. Collect the data as you go through the audit, noting the document identity, the location where collected, and anyone you discussed it with. You should plan on checking the document control system later in the audit and do the same for any records. It is a waste of time to stop your audit in a particular location to run to the information management office or records office when you can do it all at once later.

One way to collect data as objective evidence is by camera. A picture of a situation is difficult to argue with, especially an instant one, such as one by Polaroid or Kodak. Many auditors are authorized to photograph at a facility, particularly first-party auditors or second-party auditors under contract to the auditee. Be sure you have the authorization before you do and, in some facilities, you may need to have a documented permit or property pass.

The review of data collected will be used to confirm conformance with EMS requirements. The documents are to demonstrate implementation of the EMS, the environmental policy, and the objectives and targets. They are to demonstrate implementation of the commitments to continual improvement and the prevention of pollution. They are to demonstrate the commitment to comply with relevant environmental legislation and regulations.

Once the collected data has been reviewed and a determination of conformance made by the auditors, the results will be discussed in the daily team meetings. Each auditor will give a sort of "mini" report on the day's activities, data collected, interviews, and findings from review of objective evidence. Significant observations will be discussed.

The collection of data as objective evidence may take more time than calculated in the audit schedule. There is an opportunity at the daily team meeting to address this and make necessary changes. Each auditor will recommend which items they consider should be written up on what they have found. It may be that what one auditor found goes beyond his or her scope within the audit and should be looked into by an auditor evaluating that subject.

Findings should be written daily as the audit progresses. The author provides an example of an appropriate report form, the Environmental Corrective Action Request (ECAR), in Chapter 12. It is a simple format to capture your findings in a convenient and straightforward manner. Once written, the ECARs are reviewed by the lead auditor for concurrence. On certain EMS audits, legal counsel may also want to review the ECARs.

The author has found that at least 20% of your audit time will be required for report writing. Do not attempt to write the ECARs during the daily team meeting, but do so either before or afterwards. Team meetings need to flow quickly and take as little time as necessary to communicate the status of the audit that day.

Categorizing Deficiencies

The author has found categorizing nonconformances based on two thresholds of severity works quite well in collating the deficiencies for corrective and preventive action:

- A = Critical
- B = Major
- C = Minor

Start at the bottom of the categorization ladder. A "minor" nonconformance, Category C deficiency, is either (1) when a single nonconformance is found, or (2) when an observation or concern is noted.

A "major" nonconformance, Category B deficiency, includes those in which a significant number of minor nonconformances occur in one requirement.

A "critical" nonconformance, Category C, applies when (1) the total requirement or a significant part of the requirement of any given ISO 14001 criterion is deficient, or (2) there is a single critical EMS nonconformance.

Chapter 12, EMS Audit Reporting, will illustrate an example of a type of EMS Audit Report and Corrective Action Request (CAR) where this information may be indicated.

Flowchart Assessment and Auditor Checklist

Flowcharts are usually uncontrolled documents, for information only. Those in the back of procedures are usually only good for auditors. Used for information only, to give a basic overview of how a system or process works, flowcharts are useful, but they lack the necessary sophistication to be used in lieu of procedures.

An EMS auditor needs to ask the auditee whether any flowchart found during the audit is a controlled document. Is there no formal procedure readily available? Does the flowchart indicate HOW the tasks will be accomplished? Does it indicate key roles, responsibilities and authorities? Does it reference applicable forms/records for documenting objective evidence?

If the auditee is performing EMS functions to a flowchart, ask them the big question: What happens if...?

For small companies who are new to formal procedures, the use of flowcharts is a good indicator that they are trying to get control of their processes, but they cannot stop there. Procedures are required by ISO 14001, not flowcharts. The author has been to several facilities where the flowchart was the only document in use. Needless to say, the effectiveness of the EMS was less than adequate.

Formal, controlled procedures allow an organization to take credit for how it does something. Procedures result in a consistency in methods used, in the time involved, and definitely, in the final outcome, activity, or process.

Let's go back to 4.4.4 of ISO 14001, EMS programme. It says there shall be information in paper or electronic form that describes the core elements of the EMS. What are more at the core of an EMS than procedures? Information in either of the two medias is documentation. So it may be that any procedure required to satisfy ISO 14001 requirements should be documented. Documented means formalized in this context.

What do flowcharts tell you? What takes place, but not how or who or what precautions to take, special equipment needed, or when to ask for help. They are simplified diagrams of the what. Most flowcharts are in steps or block diagrams. Few call out the exact forms to use, or the current revision, or where to get them. Few call out the specific records to generate.

Standard Assessment Questions

There are certain EMS audit questions that can be asked no matter which of the 17 criteria you are auditing. Keep these questions in your auditor's tool box for use anytime you need or want to evaluate the interrelation of system processes.

- Does the procedure conform to the requirements?
- Are document control measures in accordance with ISO 14001, 4.4.5?
- Are responsibilities, authority, and roles clearly defined and documented? (4.4.1)
- Is the environmental policy available? Communicated? (4.2)
- Is there objective evidence of training on operational controls? On significant environmental aspects? Significant impacts? EMS procedures? Emergency preparedness and response procedures? (4.4.2)
- Review the availability and use of environmental records in accordance with the records management procedure. (4.5.3)

For other EMS audit questions, see Chapter 5.

A few pointers the author has learned along the way and over the years for EMS audit checklists:

- There is no set number of questions to ask but too few or too many is inappropriate.

- Commas in a requirement divide the requirement into more than one.
- Checklist questions are the core questions and not the only ones you will ask during an EMS audit. They provide guidance.
- Ask no compliance questions, only questions on conformance to ISO 14001.

Auditor/Auditee Interfacing

The EMS auditors will interface with members of the audited organization almost constantly during the audit. Keep the relationship cordial and courteous throughout. There are any number of personalities you may encounter during an audit. In any case, the auditor is expected to remain "The Professional" and leave personal likes or dislikes behind in the hotel closet. Most auditees will go out of their way to assure that you have the interfaces you need, when you need them. At times, the team leader may need to coordinate a little extra effort by the auditor or auditee.

Data to Record

Record sufficient information to support your audit report. Note the document by its identity and number, the revision and date, the location where collected, who it was discussed with, and identify your observation with regard to conformance to EMS requirements. When you evaluate document control, check for the proper review and approval of the data, if appropriate. Make note of the type of medium (paper or other form).

You are looking for data as objective evidence to demonstrate the EMS has been established and is maintained, is implemented, is periodically reviewed for suitability and adequacy, is revised when necessary, audited, and has its performance evaluated.

Types of data to review and record may include:

1. **Environmental policy**
   - Evidence of an environmental policy
   - Evidence of internal policy distribution
   - Evidence the policy is available to the public

2. **Environmental aspects**
   - Evidence of identification
   - Evidence of significance screening
   - Evidence of significance determination

3. Legal and other requirements
- Requirement identification data
- Evidence of access to requirements

4. Objectives and targets
- Environmental objectives data at each level/function
- Environmental targets data at each level/function
- Evidence of review of objectives
- Views of interested parties data

5. Environmental management programme
- Evidence responsibilities have been designated at each level/ function
- Evidence of resource allocation for means and time frame
- Evidence of review of program for necessary amendments

6. Structure and responsibility
- Organizational structure data
- EMS management responsibility data
- Employee EMS function assignment data
- EMS steering committee data (for management review, where applicable)

7. Training, awareness, and competence
- Evidence of identified training needs
- Evidence of procedures for awareness at each level/function
- Evidence of competence determination

8. Communication
- Evidence of internal communications
- Evidence of external communications
- Record of decision on communicating externally

9. EMS documentation
- EMS core element data
- EMS interactions data
- Evidence of direction to relevant documents

10. Document control
- Evidence of control of documents
- Verification documents identified

- Verification documents can be located
- Evidence of periodic review and approval
- Evidence of legibility
- Evidence of control procedures
- Evidence of procedures for creation/modification of documents

11. Operational control

- Evidence of significant environmental impact associated operations identified
- Evidence of operational conditions specified
- Documented procedures
- Documented operating criteria
- Evidence of procedures for goods/services with significant environmental aspects
- Evidence of procedures for communicating with suppliers and contractors

12. Emergency preparedness and response

- Evidence of procedures
- Evidence of preparedness
- Evidence of response capability
- Evidence of mitigation capability
- Procedure test records
- Post-accident/emergency procedure review records

13. Monitoring and measurement

- Procedures
- Evidence of performance evaluation
- Monitoring of equipment calibration data
- Evidence of regulatory compliance assessment

14. Nonconformance and corrective and preventive action

- Evidence of procedures
- Evidence of corrective/preventive action evaluation for adequacy
- Record of changes to documented procedures from corrective/preventive action

15. Records
- Evidence of procedures
- Training records
- Audit results
- Evidence of proper storage and maintenance
- Record of retention times

16. EMS audit
- Evidence of audit programme/procedures
- Evidence of audit schedule, reports and follow up

17. Management review
- Evidence of information collection for EMS suitability, adequacy, and effectiveness evaluation
- Review documents

The Classic "Time-Wasters"

Here are some of the classic time-wasters the author has found common to audits of all types and sizes:

- Auditors have insufficient time
- Long and unnecessary pre-audit meeting
- No plan of attack
- Lengthy presentations
- Unavailable interviewee
- Conducting usual business
- Touring
- Setups
- Wrong person
- Too many people

Auditors rarely have sufficient time to do all they would like to. Keep the tight audit schedule always in mind. Stick to the scope, purpose, and objectives of the audit. Follow the agenda.

Watch for time wasters, such as long and unnecessary time in the entrance briefing. This is a pre-audit conference, not a slide show on the origin and evolution of the auditee's organization. Not having a clear plan of attack is a time waster, and this is the audit team leader's fault if it occurs. Lengthy presentations by anyone—the auditee or auditor—are unnecessary.

It is a waste of time when the responsible party to be interviewed is not available at the scheduled time and location. If the auditee is interested in your completing the audit, they will see to it that interview schedules are met. An auditee conducting his or her usual business during an interview is a time waster. The interviewee should be as involved in the interview as you are.

Touring the area except during the progression of the audit, as we said before, is seldom necessary. The same is true for contrived setups for the auditor's benefit, such as mock demonstrations. You want to see the real thing as it happens.

Interviewing the wrong person can be prevented by effective planning and by gathering the correct information during the entrance on who to interview. Remember, we suggested the auditee point out the key personnel you will want to contact.

Documenting too many of the same type of adverse findings—a handful of samples, of symptoms of a problem—is sufficient. An audit is only a snapshot in time. Keep it short and sweet (new KISS principle). Stop, draw a conclusion, and move on.

Operational Verification and Assessments

An EMS audit consists of a lot of document review. It involves evaluation of the EMS administrative and management functions but also involves the observation of operational controls. This is where the rubber meets the road, so to speak. It is where environmental performance may be the most important aspect to evaluate.

In ISO 14001, there are two subclauses worth reiterating, in part, here. The first is 4.4.6, Operational control, and the second is 4.4.7, Emergency preparedness and response.

Facility operations may be normal or, at times, abnormal. The operating conditions may change from business as usual to business during an adverse incident or event. The controls need to account for both conditions.

Operational controls are to be associated with those identified operations and activities of the auditee that have the identified significant environmental aspects. These activities are to be planned, including maintenance, so they are carried out under specified conditions. Situations where their absence could lead to deviation from policy, objectives, or targets are to be covered by documented procedures. The potential for accidents and emergency situations is covered by procedures. Procedures are required for response to accidents or emergencies and for preventing and mitigating associated environmental impacts.

Procedures are to stipulate operating criteria where their absence could lead to nonconformance. In light of this, personnel responsible for implementing these procedures should be competent based on education, training, experience.

The standard requires that procedures for emergency preparedness and response be tested periodically where practicable. Responsible personnel should be competent in these procedures as well.

You will certainly be able to observe implementation of the operational controls under normal operating conditions, even if the facility is in shutdown or standby mode. It may be that you will get the opportunity to watch a response to an abnormal situation. Or you may schedule the audit to be conducted when an emergency preparedness training drill is in progress or the response procedures are being tested.

Think of operations in both situations: business as usual or in adverse conditions. The EMS is to be designed and implemented to accomplish environmental management of significant environmental aspects at all times and under any condition.

11

The Post-Audit Conference

Chairperson Guide for Conducting Exit Interview

The author believes there are just a few simple steps to an effective post-audit conference, also known as an *exit interview, exit briefing,* or *closing meeting.*

1. Thank organization
2. Distribute attendance list
3. Provide disclaimer clause
4. Request questions held
5. Allow auditors to present their own observations/findings
6. Allow lead auditor to provide impression and summary
7. Organization asks questions
8. Visit to determine corrective actions
9. Follow-up audit
10. Determine to whom draft/final reports will be sent to
11. Report forms (use, intent, purpose)
12. Organization's comments

Before you conduct the exit interview, there are some steps the audit team and audit team chairperson need to accomplish before the critique. Each auditor must share their findings with the other team members, finalize the findings, and combine findings by category of nonconformance and by any common denominator criterion of ISO 14001. Then the reports of the audit are finalized and reviewed by the lead auditor. The lead and audit team will reach a consensus on exit presentation tactics, such as which auditor will discuss which finding, what the sequence of findings to be presented will be, whether any compliance issues are to be discussed, and how much time each auditor will have.

It is important to avoid the "good cop, bad cop" style of presentation. For one, an EMS auditor is not an officer of the law or a regulator, but merely an EMS auditor with findings of conformance and/or nonconformance to report. It is even more important that you are heard during the presentation. The lead should coach the audit team on strong, but not overwhelming, oral presentation skills. The lead should remind each auditor to simply "tell it like it is." The truth is the truth, no matter what the opinion of the auditee.

Last, but in no way least, remind each auditor to just relax and enjoy the presentations. You will be on the road very soon and out of Dodge, so they say!

Now, for the guide to a successful exit interview. These seven simple steps will take you in, through, and out of the critique with any auditee, any size audience, at any location.

✪Thank the organization✪

Thank the auditee, as you would a host, for their hospitality and their assistance. Highlight any particularly noticeable actions the auditee may have taken to make your visit comfortable and efficient. Perhaps the auditee went out of their way to enable you to complete your audit. Whatever favorable deeds were performed for you, give the thanks and appreciation that are due.

✪Distribute attendance list✪

The attendance list is a record of who attended and was informed of the findings of your audit. It may be a prearranged form or a simple sheet of paper. Start it with the audit team leader, then pass it through each member of the audit team, and then route it through those attending. Make sure it has the names, titles, and phone numbers of individuals in attendance, as a minimum. And make sure their handwritten names and position titles are legible enough to understand the correct spelling. You will want to ask the auditee to run copies before you leave so they have at least one and you have the original.

✪Disclaimers clause✪

An EMS audit is a collection of random samples at a specific point in time. You will want to warrant the results of the audit by reminding the auditee that the results of the EMS audit are a snapshot in time and nothing more. Keep in mind that EMS audit results, once documented, may become prone to disclosure to lawyers, regulators, or even the public. The protection from disclosure of EMS audit reports has yet to be firmly established. Pay attention to what you put in your reports and be sensitive to the possibility that being less than precise or being careless may bring you unforgettable regrets.

✺Request for questions✺

Ask the auditee to hold any questions concerning the audit or its findings until the end of your team's presentations. Otherwise, you will find the exit interview will take hours, days, or months to finish! It is proper business meeting protocol to do this, and it fits the established rules of order for conducting a business meeting.

✺Presentations✺

The lead auditor will initiate the presentation of findings with general comments on what the findings indicate. Then, each auditor will present his or her findings and recommendations on what the auditee needs to do to resolve the nonconformance. Note what I said: recommend **WHAT,** not **HOW**. You are an auditor, not a consultant.

✺Overall impression✺

The lead auditor should summarize the findings of the audit following their presentation. This is when observations to be reported, in addition to the findings, should be identified. Give observations such as strengths and weaknesses yet to be identified to the auditee. Present concerns that required attention by management during the audit.

✺Auditee questions✺

The presentations are completed, the overall impression has been given, and now it is time to allow the auditee to ask, inquire, clarify, or confront you. The lead auditor must hold this portion of the exit to a brief time period. You and your audit team members want to tell the facts, give their thanks, and depart. It is time to go home.

✺Post-audit visit✺

If the audit team will be coming back to determine whether corrective actions are progressing adequately or to provide any assistance requested by the auditee, the visit will be informal. Nothing in ISO 14001 requires or recognizes follow-up to an EMS audit. First-party auditors will no doubt be asked to come back and help. Second-party auditors may be. Third-party auditors cannot, other than to indicate whether the corrective or, for that matter, preventive actions are close to hitting the mark with regard to the findings.

✪Follow-up audit notification✪

If you are requested to conduct a follow-up audit or are required to do so by contract, procedure, or standard operating practice, then inform the auditee that a formal letter of notification will be sent. It will be sent at least 30 days prior to the follow-up visit to allow the auditee and your own management time to arrange the resources, make assignments, and confirm the schedule.

✪Draft reports✪

Inform the auditee who will be given an opportunity to review the draft audit finding reports. Discuss what types of review are expected, by whom, for what reasons, and what will be done with any review comments.

✪Audit report forms✪

There are four basic types of audit report forms in this author's opinion: the summary report, draft report, final report, and the Environmental Communication Memo (ECM) or similar report for compliance/safety concerns.

Each of these reports is discussed in detail, and sample formats are provided in Chapter 12 on EMS audit reporting.

Attendance Lists

The EMS audit team chairperson's copy of the original or a facsimile copy of the exit-briefing attendance list should be considered a record of audit implementation. It should be retained until the next audit of the facility or until corrective actions and preventive actions for any identified nonconformance are completed.

Auditee Comments/Reaction—Disclaimers Clause

The issue of a disclaimers clause was discussed earlier in this chapter. You'll want to warrant the content of the EMS audit report but no more than that. You want it stated that no proprietary or privileged information and no confidential or security information is contained in the report, to the best of your knowledge.

Confidentialities

Allow the auditee to identify any confidentialities they may desire. You will no doubt check with your own management, or your client, and, in some cases, your legal counsel or your client's.

Communication Skills

We have provided guidance on communication skills throughout this book. For the post-audit conference, the same skills apply.

The audit team will be communicating orally and using the written audit reports. It is important that each member stick to the scope and content of their reports without meandering. The team will be in front of the auditee's representative(s), so posture, poise, and demeanor will be avenues of communication. Remaining "The Professional" consists of what you do, what you say, and how you present the results of the audit.

If an interpreter is required, speak in phrases short enough to be interpreted but without becoming dull or boring. The skills of your interpreter will determine this. Usually, an interpreter can translate about 25 words at a time.

Visual aides may or may not be effective at the post-audit conference. Also, they will require time to prepare and always lengthen the presentation time. You should figure 3 minutes per overhead or transparency, if used.

Your purpose at the exit briefing is to report the results of the audit factually, clearly, and without personal opinion, interpretation, or bias. This is an opportunity to report the facts as you found them to the auditee.

Presentation Methods

The most successful presentation methods the author has found or created include the following.

The audit team should sit collectively at a table facing the auditee's organization or at a large conference table with the auditee opposite the team. The team leader chairs the meeting. Each auditor presents their findings and observations. The team leader summarizes the overall results of the audit, invites questions from the floor, and coordinates responses to the questions. The team leader asks for statements from the auditee and manages any discussion between team members and the auditee's members.

The entire presentation must be professional, polite, and diplomatic. It must be well-organized and pre-planned. No individual presentation should exceed 20 minutes. The entire exit should be completed in as short a time as necessary in the given environment to present the audit results, provide for future or further communications, and end the visit at the facility.

The Conference Room

Where the post-audit conference is held will be at the discretion of the auditee. In light of the number of nations, organizations, and locations around the globe where EMS audits will be conducted and post-audit conferences held, the concept of a conference room will no doubt change dramatically.

The typical conference room usually has a large table for seating up to 40 people. Occasionally, the conference for a post-audit briefing will take place in more of a theater, with a podium and dais.

No matter what facility is utilized for the exit, the purpose of the conference is the same: to present the results of the audit.

Report Distribution

The distribution for the draft and final audit reports will be established in the audit procedures, contractual documents, or other planned arrangements between the auditor and auditee's organizations. As a minimum, the auditee will receive the report. For second-party and third-party audits, additional distribution will be established, such as to the auditing organization's management. The report should only be distributed to those with a need to know the results. Remember our discussion of confidentialities.

Follow-up

No follow-up to an EMS audit is required by ISO 14001. However, follow-up may be required by procedure, by contract or other binding agreement, or may be requested by the auditee or your client as auditor. Chapter 13 provides information on the follow-up audit.

Corrective Action Direction

The EMS audit reports should include recommendations for corrective action. Never try to tell an auditee *HOW* to correct a nonconformance, only *WHAT* they should accomplish to satisfy a given requirement. It is the ultimate responsibility of the audited organization to decide on the corrective and preventive actions for EMS nonconformances. They own and operate the systems.

Auditors may recommend that procedures be established and maintained, that certain data or procedures be documented, that certain activities be defined—all from the wording within ISO 14001. For example, for a failure to have procedures for internal communication, the auditor may recommend the auditee "establish and maintain procedures for internal communication between the various levels and functions of the organization." For indeterminate confidence of personnel, the auditor may recommend the auditee to determine that "personnel performing the tasks which can cause significant environmental impacts" are competent based on "appropriate education, training, and/or experience."

Remember, recommend *what* to accomplish, not *how* to do it.

12

EMS Audit Reporting

Data to Record

During the collection of objective evidence during an EMS audit, your team will want to make note of at least these few pieces of information:

- Location
- Subject
- Document numbers
- Person discussed with
- Deficiency/comment/information

Noting the location in the facility where the objective evidence was found or observed will give you pertinent information for writing your audit reports, discussed later in this chapter. Why is the location important? It is essential for verifying the evidence and for the auditee in designing corrective or preventive action.

The subject is the criterion or portion of a criterion within ISO 14001 to which the evidence applies. For example, a controlled procedure for process X found during the audit in the shift supervisor's office was incomplete and missing several pages; this is contrary to 4.4.5, Document control. Or the training records for Operator A at the waste storage drumming area were reviewed during the audit and found not to include evidence of emergency response training, contrary to 4.4.2.

Controlled documents have identities, usually, a document number, revision number, and date of revision. Often, they have titles. Whatever information distinguishes the document from any other is its identity. Recording this also aids in verifying the evidence and locating the document for further analysis and corrective action. The document may be identifiable by its traceability to an activity, item, or person.

You will want to note who you discussed your observation with. Try to record the name of the person and their title or responsibility. Actual names of people are never recorded in EMS audit reports, with only one exception discussed later in this chapter. But for verifying evidence and the potential need for the name when discussing the evidence with the auditee, it is important data to record. You will want to verify that you discussed the observation with the correct person as well.

The description of the observation, whether a deficiency, comment, or simple information, should be succinct but sufficient to fully reflect what you found. Give no opinions, only facts. You want enough detail to support a nonchallengeable audit report. This could be something said to you during an interview and should be cited with quotation marks.

EMS Audit Report Form

The EMS audit report form will be designed by the auditing organization to suit their format and style and their needs. For the purpose of illustration, the author offers the following EMS Audit Report and the ECAR forms as examples.

The ECAR has data entry spaces for:

- Organization
- Request date
- Reply due date
- Statement of requirements
- Findings/concern
- Recommendation
- Other information

The form included here is simple but allows for reporting what your EMS audit found during the visit and evaluation. It has a space for a unique ECAR identifier number and for document identity and tracking. The top portion is for administration: to whom and where the ECAR should be directed; date of the documented request; the auditing organization; a due date for the auditee to reply by; any appropriate information on a specific program or project; the criterion number from ISO 14001 or other requirement document (procedure, policy, standard subscribed to); and the location (facility). It also provides for recording any pertinent contractor document number and its effective date, who the observation was discussed with in the auditee's organization, and the deficiency category (A, B, or C).

The midsection of the ECAR is for recording and reporting the requirement, the deficiency or nonconformance found, and the auditor's recommendation for corrective action (WHAT not HOW). You should only state enough of the requirement to reflect its substance; for example, if the finding is against ISO 14001, 4.5.1, Monitoring and measurement, and the lack of documented procedures, you might write, "...documented procedures to monitor...key characteristics...." If against ISO 14001, 4.5.3-Records, for the lack of identification on a record, you might write, "procedures for the identification...environmental records."

Write only enough of the requirement to allow the reader, especially the auditee, to be able to go to the requirement for further information.

When writing the findings or concern, start with a "hammer" statement, one that drives the point straight in, such as the auditee was unable during the audit to provide objective evidence of the last emergency drill; the auditee's environmental policy document could not be located by the auditee at the time of the audit; the audit procedures fail to cover audit scope, frequency, or methodology; and results of EMS audits are not provided to management. Do you see the flavor in mind?

The recommendation, as we have said, should inform the auditee of what needs to be done to resolve the deficiency, period. You as auditor have no responsibility and insufficient knowledge about the true nature, scope, and cause of the deficiency to suggest action to correct or prevent it. All you have done is identified a deficiency. But, you do understand the requirement enough to recommend the minimum action to be taken.

Here are some typical ECAR recommendations for your consideration.

A deficiency is found against ISO 14001, 4.4.6, Operational control. The nonconformance is the lack of a documented procedure related to an identified environmental aspect of the auditee's operations. Let us say there is a fuel oil tank that has an overflow pipe. The pipe is flexible and can be routed to a waste container for the overflow, to a sanitary drain, or to a 55-gallon drum for reuse. The procedure fails to provide instructions on how the overflow piping is to be routed during normal operations or who is responsible to verify correct configuration. You, as EMS auditor, find the piping, in this case a 2-inch rubber hose running from the overflow tap to the sanitary drain. You may want to recommend the auditee establish and maintain documented procedures to cover situations where their absence could lead to deviations from environmental objectives and targets.

Let us say the auditee was unable during the audit to provide objective evidence that they had ever tested their emergency response procedure. The procedure you have reviewed provides for putting out an electrical fire in the battery room. You might recommend the auditee periodically test the emergency response procedure where practicable.

Or, let us say your audit finds that the auditee has no document control procedures. You might recommend the auditee establish and maintain procedures for controlling all documents required by ISO 14001 to ensure they can be located. You would recommend something like, "the organization should establish and maintain procedures for controlling all documents required by ISO 14001 to ensure that they can be located."

The ECAR has a second page, just in case you need more space.

Final ❑ Draft ❑	No._____
EMS AUDIT REPORT AND CAR	

To: (organization name, address, phone #)   **Request**
                                                                                    Date

From: **Reply**
                                                                                    Due
                                                                                    Date

Programme/Project	Req./No./Para.	Location

Contractor Doc. No.	Effective Date	Discussed with	Deficiency Category

Statement of Requirements

Findings/Concern

Recommendation

Acknowledgment of Distribution		EMS Audit Department	
Distributed to:	Date:	Prepared by:	
Department:	Bldg.:	Lead Auditor:	

Responsibility for Corrective Actions		Reviewed/Approval	
Name:	Title:	Name:	
Department:	Bldg.:	Title:	
Approved by:	Date:	Date:	
Title:			

EMS AUDIT REPORT AND CAR

Continuation Sheet Page of

Findings/Concern (continued)

Reporting Examples

- Summary report
- Draft report
- Final report
- ECM

Summary Report

The summary report is the one to be sent for review and comment to gather clarification information, but not to change the intent or basic content of the audit results. An example of such a summary report format is included below for consideration. This is the report the audit team may want to give the auditee at the exit briefing.

The summary report is the forerunner of the final audit report. It provides a plethora of information on the audit as conducted.

- EMS audit number and date
- Organization or departments audited
- EMS audit personnel
- Audit scope and purpose with objectives
- Key auditee personnel contacted (other than escorts)
- Identified EMS nonconformances (brief description)
- Concerns (beyond the nonconformances)
- Proposed distribution for draft and final report
- Statement of confidentiality (if needed)

EMS AUDIT **Summary Report**	Doc No. EMS-81 Rev. C 6/24/96 Rev.App._____
EMS Audit No.	**Date**
Organization/Department(s)	
EMS Audit Personnel	
Scope/Purpose	
Key Personnel Contacted	

	Title:
	Title:
	Title:
EMS Nonconformances	
Concerns	
Draft/Final Report Distribution List	
Statement of Confidentiality	
EMS Lead Auditor	**Date**
EMS Auditor	**Date**
EMS Auditor	**Date**
Acknowledgment of Receipt	
Name: Title:	Date:
Name: Title:	Date:

Third-Party Summary Report

For third-party audit summary reports, the author proposes some sort of format similar to the example to follow. It identifies the audit team's overall rating or EMS impression (acceptable, conditional, unacceptable). It allows for identifying areas of strength and weakness. There is space for comments on policy and procedures and comment on the quality of management's attitude and that of the rest of the organization.

Many third-party auditors like to provide the auditee the draft report at the exit briefing. Otherwise, a summary report may come in useful.

EMS Audit Summary Report
(Lead Auditor Summary Report)
(Third-Party Audit)

1. **What is your rating for this activity?**
    - ❏ **Acceptable**
    - ❏ **Conditional, adequate for P.O. after corrective action**
    - ❏ **Conditional, audit after corrective action**
    - ❏ **Unacceptable**

2. **In which areas or criteria do you find this facility strong?**

A)	D)
B)	E)
C)	F)

3. **In which areas or criteria do you find them weak?**

A)	D)
B)	E)
C)	F)

4. **Comments on environmental management policies and procedures reviewed during this audit:**

5. **Comment on management quality attitude and informal organization:**

Date	Signature
	EMS Lead Auditor

Report Response

The author has provided a sample type of draft report response form (below) and an example of a corrective and preventive action plan for response to the final EMS audit reports.

The draft report response from the auditee may be prepared prior to the audit team leaving the facility or soon after and submitted for consideration. It is simple in structure and content: report number, date, response to the specific nonconformance, and the conclusion of the auditee.

This is a chance for the auditee to propose any interim corrective or preventive actions to the nonconformance. It will be of assistance to the audit team in finalizing their reports and in the eventual evaluation of formally proposed actions by the auditee.

EMS Audit Draft Report Response	
Report No.	**Date**
Response:	
Conclusion:	
Title	**Date**
Title	**Date**

Environmental Communications Memo

What if, during an EMS audit, you encounter an environmental regulatory issue or a health or safety concern? Compliance, health, and safety are beyond the scope of an EMS audit. What do you do?

The author realizes that we, as ethical and professional auditors with knowledge of environmental laws and other regulations, will be confronted with issues that cannot go unreported. But, they can be reported separate from the EMS audit results. With this in mind, the author suggests something like the ECM. It is a way to identify and document your concern, notify the responsible parties of the concern, and leave the matter outside of the EMS audit.

On our ECM (next page), you will find data entries for the EMS audit number and date of the observation, as well as the organization or department and location where the observation was made. Our form identifies the EMS lead auditor, as well as the auditee contact. Then, we provide space to describe the concern. Remember, this is not an EMS nonconformance, but only a concern requiring action above and beyond the EMS audit.

Our example ECM also allows for identifying who issued the ECM and when and space for the auditee or whoever else acknowledges receipt of the concern and when. The bottom of the form, for review and reporting, is the responsibility of the recipient organization.

This form would normally be developed at the time of the observation, or as soon as possible thereafter, and delivered to the auditee's representative that same day. It may be that one copy will go to your own legal counsel and one will be for your files.

Confidentialities

An associate of the author, Fernando Rodriguez, has alerted me to the issue of confidentiality from the perspective of the most-used common law privilege in the United States, the attorney-client privilege. It is intended to ensure communications between attorney and client are complete, so the attorney can provide the best advice possible to the client. These communications must be made, of course, for the purpose of providing legal advice to a client. If the client shares information with anyone who does not have a need to know it, the client waives the attorney-client privilege.

EMS audits, to be protected by this privilege, need to be carefully structured and probably initiated by top management for the purpose of obtaining legal advice. If so, then legal counsel should most likely manage the audit, in one capacity or another. The results would then be distributed on a need-to-know basis. Whether this type of audit will sustain interpretation and remain protected from disclosure remains to be seen. The EMS audit is a review of management systems, not of compliance. The basis for the ECM is to provide

Environmental Communications -MEMO-		
		Memo No.
EMS Audit No.:	**Date of Observation:**	
Organization/Department:	**Location:**	
EMS Lead Auditor:	**Contact:**	
Concern:		
Issued by: Date	**Acknowledgment:** Date	**Acknowledgment:** Date

Regulatory Compliance Review/Report

Person Contacted:	**Actual Audit Date:**	
Findings/Concerns:		

Is Corrective Action required to be issued against this review?   ❏ YES ❏ NO

Prepared by: Date	**Approved:** Date	**Approved:** Date

Environmental Communications -MEMO-		
Memo No.	Continuation Sheet	Page of
Concerns (continued):		

information to the appropriate regulatory compliance audit function who has the responsibility and authority to examine and evaluate the potential compliance issue formally.

If the auditee shares information on its environmental impacts with a second-party or third-party auditor, whether a lawyer or not, it touches on potential liability for noncompliance. Doing so may be construed as sharing confidential information with someone who has no need to know, with regard to providing legal advice. If so, there can be no attorney-client communications privilege.

Corrective/Preventive Action Reply

The final response from the auditee may be presented in a form similar to the corrective and preventive action plan. Again, the form shows the report number; date; the results of root cause analysis, if any was conducted; any interim actions taken, whether permanent or preventive; the corrective actions proposed or taken; the preventive actions proposed or taken; and when the actions become effective or when completion can be expected.

EMS Audit Corrective and Preventive Action Plan	
Report No. **Date**	
Root Cause Analysis:	
Interim Actions:	
Corrective Action:	
Preventive Action:	
Effective Date **Est. Date of Completion**	
Signature of personnel responsible for corrective/preventive action:	
Prepared by	**Date**
Approved by	**Date**

You may get an unacceptable auditee response, "I didn't have the time, the money, the resources...." These point to system problems and are not reasons, but excuses. They indicate a deeper, more serious management problem. If you receive this type of response, take action. Request a new response from the auditee.

Types of Reports

You, as EMS auditor and, particularly, as audit team chairperson/lead auditor, will be concerned with all the types of audit reports discussed earlier in this chapter. You will draft your findings, review the draft responses for acceptability (is the proposed action in the ball park?), then finalize and issue the final reports. Your ECMs, if any were generated, should be handled only once, when you first document your concern and deliver it to the responsible party or parties. Note that the ECM fits nicely into the expectations of ISO 14001 4.4.3, Communication.

Words to Avoid

- Never/no
- Must
- Always
- Shall
- All
- Direct/specific method

These words will not get you the kind of response from an auditee you are expecting. Use of "never" is like the use of "all"—it is too universal, too large in scope. And your audit was conducted over a few days, so how can you possibly say "never"? Same for "no," with certain exceptions and in certain contexts where "no" may be used positively: "No record of the last calibration was available at the time of the audit," or "significant findings were identified by the audit."

"Must" and "shall" are requirement words, not words to put in findings or recommendations. They establish criteria to be met. You have identified deficiencies in meeting specified criteria, and your report should be written to proactively seek proper attention by the auditee, with the proper corrective and preventive actions.

"Always" is a long time. Can you possibly even think in eternal times during a few days of EMS audit?

Never cite or recommend a direct or specific method. That is telling the auditee how to do something and is not your job or your responsibility.

Avoid using "etc.," as we discussed earlier in this book. Instead, use "e.g." or "for examples of" for the full scope. Avoid "and/or" (ambiguous); use one or the other, depending on the context (and...or). Use "we" (for the audit team) or "the audit team." Do not use personal pronouns like "I."

Words to consider using instead:

- Not always
- Recommend that
- Findings indicate
- Develop, document, implement, and maintain
- Determine magnitude
- Review and determine

Avoid the numbers game, as mentioned earlier in this book. It serves no useful purpose and is reminiscent of earlier business days when bean counting was the latest state of the art, even though ineffective and even damaging. What do these phrases tell anyone?

3 out of 15

2 out of 30

1 out of 12

Challengeable/Nonchallengeable Reports—Recommendations

The one thing you want least as an EMS auditor is to have your reports challenged. Here are some suggestions of challengeable and nonchallengeable observations:

- **Challengeable observation**

  It was observed that there were no procedures for operators to dispose of by-product material.

- **Nonchallengeable observation**

  Observations indicate a lack of procedures available for disposal of the following by-product materials (i.e., identified significant aspects):

  1. Hydrogen-3 (tritium)   Control No. WD 64279
  2. Carbon-14   Control No. WD 64972
  3. Solidification Agent   Control No. WD 67429

Avoid these types of words in your audit reports:

- Comply with the procedure
- In accordance with...

- Fix...

- Procedure X was Revision C but should be revision D; purge the system

- Comply with the specification

Comply? Wait. This is a conformance assessment and not a compliance inspection or compliance audit.

In accordance with? This is a "how," not a "what."

"Fix" excludes preventive actions. Avoid this one like the plague. Fixing an identified problem without root cause analysis, if it is significant, will probably only serve as a band-aid, and the problem will recur. Root cause analysis is a highly specialized skill requiring training and experience, like leading an audit.

As for challengeable recommendations:

- **Challengeable/Inadequate Recommendation**

  All by-product material must have procedures.

  OR

  Prepare inspection, handling, and storage instructions.

- **Nonchallengeable Recommendation**

  Determine the extent to which inspection, handling, and storage procedures have not been developed. Determine steps necessary to ensure procedures are generated for the control of by-product materials.

  OR

  Review and determine actions necessary to ensure by-product material is controlled in accordance with documented procedures.

Summary

This brings us to Phase III of the EMS audit. This phase has four basic steps:

1. Issue draft report
2. Issue final report
3. Provide action plan
4. Provide follow up

❂Draft Report❂

- Provide corrected closing report
- Determine distribution list
- Distribute draft report
- Allow time for correction/clarification

❂Final Report❂

- Provide corrected draft report
- Highlight requirement for action plan
- Determine action plan preparation deadline

❂Action Plan Preparation and Implementation❂

- Based on audit findings in final report

❂Follow up on action plan and close-out❂

This final step is covered in the following chapter.

Distribution

The distribution of the EMS audit report at any stage of development and finalization will be determined by the auditor, client, and auditee. The auditee will be the party to whom the report is sent for response and corrective/preventive action. The auditor's management most likely will be on distribution. Nothing in ISO 14001 requires anyone to receive the audit results except for the auditee's top management for management review under 4.6.

13 Follow-Up Audit and Close-Out

Audit Report Collection

No audit follow up is required by ISO 14001. The follow up for nonconformances is the responsibility of management for continual improvement. But no audit is complete without some type of follow up on actions taken to resolve identified deficiencies. No audit procedure has yet been written without requiring follow up and close-out of audit findings.

The standard practice of audit involves revisiting the identified deficiencies and nonconformances and the effectiveness of their correction and prevention from recurring. As you have learned in this book regarding EMS auditing and conformity assessment, the report forms and response forms concern analysis of the nonconformances, and corrective and preventive actions.

Once you have completed an audit and have submitted your audit report to the auditee, they respond with a proposed action plan. You evaluate the proposal against the identified deficiencies and, once satisfied, the auditee implements the plan. Now it is time to return to the auditee's facility and verify the action was taken as planned and that it corrected the deficiency and prevented its recurrence if a significant condition adverse to environmental management or performance existed. Sometimes, midway into a corrective action, the auditee will find perhaps a more effective action than originally planned. That is acceptable. It is a conscious management decision.

Your verification of corrective and preventive actions will involve having copies of the applicable audit reports identifying the deficiencies and nonconformances. Any other pertinent data should also accompany you on your visit. Most audit procedures call for the auditor who identified the nonconformance to verify corrective action. This is not always practicable or necessary. If the audit was done professionally, the results will be reproducible, and any competent auditor will come to the same conclusions, based on the objective evidence. The author has evaluated corrective actions for audits by others and has informed the auditee that the actions appear satisfactory. This is further supported by the author finding no objective evidence of any similar or the same problem.

Review

In a follow-up audit, the auditor reviews objective evidence of the actions taken to correct or prevent the problem. This may include a tour of the area where the nonconformance was found. It may include interviews of responsible personnel. In essence, it is an audit of as-found conditions just as it was in the original audit. But this time, the purpose is to validate effective actions have been completed.

Focus should be on the nonconformance as identified, the actions as planned, and the actions as taken in review. The scope, purpose, and objectives of the follow-up audit is to confirm a return of the affected system to conform with requirements.

Document/Procedure Assessment

Most EMS nonconformances will involve a procedure. Sometimes a procedure will need to be developed. Sometimes one will need to be revised. You should review the original documents, if any, against the content of the new ones to compare the differences. Other times, the corrective or preventive action may involve generating an EMS document. Perhaps a job description, program plan, lesson plan, or other document needs to be completed. Assess the contents, distribution, and use of the document or procedure. A document is only as good as it is implemented.

Change Control Verification

Your follow-up assessment is a verification of the changes made to the deficient management controls. It is also a verification of the control afforded the changes. A change will require responsible personnel who are aware of the change, trained on significant changes, and competent to implement the change determined by management. Changes should be evaluated by management review for their impact on the suitability and effectiveness of the EMS.

Changes to documented procedures are required to be implemented and recorded by 4.5.2 of ISO 14001, Nonconformance and corrective and preventive action. It is very likely correction or prevention of the identified nonconformance will result in changes to documented procedures. Your follow-up assessment, then, should be prepared for a check of the records.

Corrective and preventive actions for EMS nonconformances are not simple fixes. They have a domino effect throughout the EMS. The higher the tier of the change in the hierarchy of EMS documents, the greater their impact on the systems. A change to environmental policy, objectives, or targets will require major changes in the means of implementation. A change to an EMS procedure or document will have repercussions down through operational controls and records, and so on down the chain of control.

Obsolete Document Controls

Changes made to EMS documents or procedures as a result of corrective or preventive action will make the prior versions of the documents or procedures obsolete. As you know, ISO 14001 4.4.5-Document control, requires obsolete documents to be promptly removed from all points of issue and points of use. An alternative is to mark them in a suitable manner to preclude their inadvertent or unintended use. If these obsolete documents are retained by the auditee for legal purposes or the preservation of knowledge, they must be identified in a similar manner.

Your follow-up assessment will involve verifying the control of obsolete or outdated documents.

Corrective Action

Corrective action is to eliminate the cause of the actual nonconformance. It must be appropriate to the magnitude of the problem. It must be commensurate with any environmental impact encountered. It is corrective action, not mitigating action.

To be effective, the corrective action will be based on analysis of the cause or causes of the nonconformance. If a significant nonconformance (Category A), root cause analysis should be performed by the auditee in developing the corrective action. Corrective actions are carefully planned, scheduled, implemented, and evaluated for success by the auditee. You, as auditor, will verify these management activities have been accomplished and have eliminated the cause or causes of the nonconformance.

Preventive Action

Preventive actions are taken to eliminate the potential of a nonconformance. They, like corrective actions, must be appropriate to the magnitude of the potential problem and commensurate with any potential environmental impact that may be encountered.

Prevention is one of the most relevant tools of an EMS. It is an indicator of the commitment to continual improvement in environmental performance. Prevention is a form of liability insurance as well, and it is fundamental to Total Quality Management.

Your assessment of preventive actions taken is a little more speculative than assessing corrective actions. But the requirements are clear, and actions taken to prevent nonconformance will be known to be effective or not be effective the first time they are called upon to work. For instance, the auditee revises a procedure to prevent a potential nonconformance, but the deficiency occurs anyway. Was the preventive action appropriate to the magnitude of the problem? Was it commensurate with the potential environmental impact?

Preventive actions for an EMS will most often be very programmatic in their scope.

Verification/Validation of Change

As we discussed in this chapter, the primary function of an EMS follow-up audit or assessment is to verify that change has been made and validate the effectiveness of the change. Observe changes to activities in action, if you can. Interview responsible personnel at every level and function affected by the change as to their awareness and understanding of the change and the need for the change.

If the change involves a change in personnel, you are interested not in the individuals involved but in the effectiveness of the reassignment of roles, responsibilities, and authority.

Follow-up Audits

Follow-up EMS audits will be planned around the original audit's scope, purpose, and objectives. They are re-audits, so to speak. Their intent is to verify that corrective or preventive actions taken have been completed or implemented and address the full magnitude of the problem, commensurate with the significance of any actual or potential environmental impact.

Reporting

The results of an EMS follow-up audit or assessment will be reported using the same distribution as for the audit report unless there have been changes in responsibilities. The auditee will want to be notified of your follow-up findings or any new concerns you may have identified. The auditor's management will want a record of your corrective and preventive action evaluations. A copy of the follow-up report should be filed with the audit report.

Future EMS audits of the same areas where a nonconformance was identified may want to include the corrective/preventive action in their scope.

Close-out

An EMS audit may be closed out when all actions, whether corrective or preventive, have been taken and their effectiveness evaluated and found satisfactory. When the actions taken are found less than satisfactory, the identified issue remains open. Open issues should be included in the scope of future EMS audits as priority items.

Third-party audits will remain open until all actions are completed and verified. Many second-party audits will also remain open for the same reasons. A first-party audit may be closed out by capturing the audit findings in a formal corrective action system.

No corrective or preventive action should require more than 60 to 90 days to be fully completed or implemented. In fact, most actions can be accomplished within 30 days. If the action requires longer, it should be thoroughly questioned.

Follow-on Assessments

Follow-on assessments are evaluations of corrective actions in progress to observe their effectiveness as they are completed. This type of assessment works very well with evaluating the potential effectiveness of preventive actions. Ask the "what if" questions and the "if...then" questions.

First-party EMS audits may want to follow on to second-party or third-party audits. If outside organizations found nonconformances, then the auditee's own auditing organization should ensure they are corrected and prevented.

Training/Qualification for Change

As we said earlier in this chapter, any change to any element of an EMS will require changes in the training, awareness, or competence of responsible personnel. The old dog must learn new tricks, so to speak, for the EMS to work effectively and continually improve an organization's environmental performance.

The amount and depth of changes to training or qualifications will be directly proportional to the depth and breadth of the change to the EMS. Again, the hierarchy of the change is a determining factor. Change to environmental policy, objectives, or targets will affect all personnel. Change to an EMS program procedure will affect most personnel. Change to an EMS operation or activity will affect personnel responsible for the operation, the activity, control of the operation/activity, or the verification/validation of the operation or activity. Change to an EMS record may require no change to training or qualifications.

14 Supportive Environmental Audits and Assessments

Reasons for Conducting Environmental Audits and Assessments

The author can think of ten distinct and different reasons for conducting an environmental audit or assessment:

- To ensure compliance
- To define liabilities
- To protect officials
- To investigate acquisitions/mergers
- To track compliance costs
- To transfer information
- To provide management accountability
- To provide insurance information
- To train
- To respond to Security Exchange Council (SEC)

One of the main, and let me say, apparently, most popular, reasons for an environmental audit is to assess and ensure compliance. The status of compliance is reported to management, as well as to interested parties such as the regulator. A second reason is to define liabilities, which may be accomplished by risk assessment, as discussed later in this chapter.

As time goes on and more organizations come under regulatory scrutiny and as more nations begin to "enjoy" environmental regulations, more environmental audits will be conducted to identify liabilities and potential improprieties, and to protect company officials.

An environmental audit may be called for in an acquisition or merger or requested by an insurer. In fact, a comprehensive audit of your environmental state and recognition of known or suspect problem areas may result in reduced insurance premiums. Phase I site assessments are often required for innocent landowner defense.

Tracking compliance costs may be another reason for an environmental audit.

An environmental audit may be performed to share information between facilities or sites of an organization. This may provide a form of benchmarking with regard to environmental compliance performance or to share lessons learned at one location with others.

An environmental audit may provide more management accountability with regard to compliance or performance. It is a viable way to generate an analysis of the gap between what is being managed and what needs to be managed. It is also a way to verify or validate the management oversight process.

Training personnel on environmental regulatory compliance or environmental performance, and on how to identify significant environmental aspects, is another justification for an environmental audit.

The U.S. Securities Exchange Commission, or equivalent organization in other countries, may request an environmental audit for the company's profile rating on the stock market.

There are six basic types of environmental audit:

- Type 1: Compliance
- Type 2: Effectiveness
- Type 3: Risk assessment
- Type 4: Environmental Impact
- Type 5: Life-Cycle
- Type 6: Remediation, Restoration

The results of these six types of environmental audits will usually bring forth a plethora of related inspections, assessments, and studies to further support the extensive need for information. Many of these are not a means to an end result, but rather an extensive ongoing study to better deal with a regulated environmental impact and management's continuous efforts to deal with change.

Type 1 audits may be performed to verify compliance to operational procedures (EMS, occupational health and safety, quality assurance) as well as to other regulations (local, civic, state, tribal, federal, agency). In Europe, under the Environmental Management Audit Regulation, there are environmental statement audits.

Role of Attorney

For the most part, the author has found that at least 40% of EMS auditors have direct access to legal counsel. Therefore, the service that may be provided by having legal counsel on an EMS audit team could be insignificant, but legal counsel may play some important roles in the EMS process:

1. Help plan the audit
2. Assess operational compliance
3. Provide guidance
4. Review
5. Advise
6. Interpret
7. Provide legal evaluation
8. Identify high priority discrepancies

The author wants to emphasize that EMS assessments should require no legal counsel beyond the scope, purpose, and intent of the audit. But it is critically important that EMS auditors have access to legal counsel.

Confidentiality is an issue we have discussed previously and want to discuss again now. ISO 14001, 4.5.4-EMS audit, requires that information on the results of audits be provided to management. No other dissemination of the information is required. However, if you have involved an attorney as part of the EMS audit team, you may want to assign counsel to audit 4.5.4 and 4.2-Environmental policy.

My colleague with a unique sense of humor has a friend who is an attorney somewhere in New Jersey, within sight of the World Trade Center. He asked his attorney friend what was the most important thing to know about the law. The attorney answered, "First, obey. Second, don't get caught."

If your EMS audit has identified any concern beyond the scope of the EMS, such as a regulatory compliance or health and safety issue, your attorney should be so informed. If you generated an ECM, make sure they are given a copy. It is protection to your favor, should the need arise.

Regulatory Compliance Audits

Environmental Legislation

An EMS auditor must have a familiarity with the country's environmental legislation of the country where the audit is conducted. This book will describe some of the more important environmental legislation of the United States. But in any country, including the United States, the EMS auditor must be familiar with its legislation directly applicable to environmental protection and performance. This legislation may be from the federal, state, regional, or local level.

The author's advanced EMS course is taught in a number of countries, each having its own environmental legislation. No two countries share the legislation, although many attempt to imitate the regulations of the U.S. Environmental Protection Agency where practicable. The legislation often changes. In most countries, the legislation improves as time goes by. Our course in Japan discusses applicable legislation there; our course in Thailand, that country's legislation, and so on.

Here is a brief overview of U.S. environmental legislation and the associated document numbers. The first number, such as 40, stands for Title; CFR, for Code of Federal Regulations; and the number following CFR, for the part of the regulation.

CAA	Clean Air Act
CERCLA	Comprehensive Environmental Response, Compensation, and Liability Act
CWA	Clean Water Act
DOT	Department of Transportation Regulations
FIFRA	Federal Insecticide, Fungicide, and Rodenticide Act
HMTA	Hazardous Material Transportation Act
NEPA	National Environmental Policy Act
OHSA	Occupational Health and Safety Act
PPA	Pollution Prevention Act
RCRA	Resource Conservation and Recovery Act
SARA	Superfund Amendments and Reauthorization Act
SDWA	Safe Drinking Water Act
TSCA	Toxic Substances Control Act

Keep in mind the new Technology Improvement Act, signed into law in March of 1996 by President Clinton, a definite promoter of accepting standards such as the ISO 14000 series.

Clean Air Act

The Clean Air Act (CAA) was enacted in 1970, then amended in 1971, 1977, and 1990. It is the federal statute mandating the prevention and control of air pollution from both stationary and mobile sources. It requires the U.S. Environmental Protection Agency (EPA) to establish three types of national standards: National Ambient Air Quality Standards (NAAQSs); New Source Performance Standards (NSPSs); and National Emission Standards for Hazardous Air Pollutants (NESHAPS).

Where air quality is better than required by the NAAQSs, there are Prevention of Significant Deterioration (PSD) regulations that establish strict preconstruction guidelines and monitoring requirements.

The CAA specifies that states must enforce the pollution abatement requirements for existing and new sources. States are free to develop more stringent regulations but cannot lower federal standards. Any facility emitting air pollutants must be familiar with their applicable state regulations.

The CAA identifies Best Management Practices (BMPs), as do other federal regulations. These BMPs are typically evaluated when statutes or regulations do not exist, do not apply, do not describe specification actions that may be required to achieve the statutory or regulatory goals, or do not go far enough to achieve protection of human health and the environment. Most BMPs complement existing regulations and provide an added degree of protection when prudently employed. When utilized, BMPs are normally described in the appropriate site operations documents of the auditee.

Title 40 CFR Part 58 is of particular interest to an EMS auditor. It provides guidance on monitoring criteria and for PSD monitoring. Title 40 CFR Part 61 is also of interest and specifies notification and reporting requirements for hazardous air pollutants. It should be noted that the 1990 amendments to the CAA may require implementation of transportation controls and clean fuel vehicle programs. These amendments also require that states implement an operating permit program, which may require modeling analyses and monitoring of sources.

Clean Water Act and Safe Drinking Water Act

The Safe Drinking Water Act (SDWA) was enacted in 1974 to establish primary drinking water regulations applicable to public water systems. The regulations specify contaminants that may have an adverse health effect and set maximum contaminant levels. The SDWA also establishes secondary drinking water regulations for contaminants that may adversely affect odor or appearance of the water.

Most states have EPA approval to administer their own drinking water compliance programs.

The general approach to the surface and drinking water portion of an environmental compliance assessment or audit includes (1) review of documentation, including the Spill Prevention, Control, and Countermeasures (SPCC) Plans, BMP and other plans, permits and permit applications, standard operating procedures (SOPs), training, operation, and maintenance records, and earlier audit and assessment reports; (2) observation of site conditions, site operations, and routine procedures; and (3) interviews with site personnel and regulators.

The Federal Water Pollution Control Act was enacted in 1956 and amended entirely in 1972 and was again amended by the Clean Water Act (CWA) and reauthorized in 1987. The intent of the legislation is to restore and protect the integrity of the nation's waters by controlling discharges of pollutants. The CWA regulates waste water discharge directly to navigable or surface waters and those indirect discharges to publicly owned treatment works. It establishes the National Pollutant Discharge Elimination System (NPDES), prohibiting the direct discharge of a pollutant from a point source into U.S. waters except by special permit. It also has regulations for SPCC Plans.

Title 40 CFR Part 123 contains requirements a state must meet in order to administer its own NPDES program.

With regard to BMPs, the NPDES permit may require facilities to develop formal BMP plans for controlling toxic and hazardous materials, whether raw materials awaiting use or waste products awaiting treatment and disposal.

Comprehensive Environmental Response, Compensation, and Liability Act, and Superfund Amendments and Reauthorization Act

The Comprehensive Environmental Response, Compensation, and Liability Act (CERCLA) was passed in 1980, as was the Superfund Amendments and Reauthorization Act (SARA). CERCLA provides authority and source of funding for cleaning up abandoned/uncontrolled hazardous materials released into the environment. SARA defines who is liable to pay for cleanup of contamination caused by past activities. The Innocent Landowner Defense Amendments of 1989, 1990, 1991, 1992, and 1993 (at last count! I told you about legislation changes!) resulted in Environmental Site Assessment mechanisms: Phase 1, a historical search into previous ownership and uses; Phase 2, site sample collection and analysis; and Phase 3, findings of Phases 1 and 2 and projection of movement of chemicals.

An environmental assessment or audit may consist of (1) review of available documentation (agreements, court orders/decrees, program plans, site action plans, federal and state documents, results of previous audits/assessments/inspections, and inventories of areas requiring cleanup); (2) examination of activities under way (with field visits where work is occurring or will occur in the future); and (3) interviews. Audits of inactive waste sites will need to address identifying and understanding specific regulatory authorities for the site response program, understanding the unique factors imposed as a result of agreements or court orders/decrees, and evaluating how the site has moved from complying only with the broad requirements of statutes and regulations to establishing detailed program plans and site response plans that reflect appropriate site-specific decision making.

The Emergency Planning and Community Right-to-Know Act (EPCRA) is known as SARA Title III and imposes requirements for providing emergency hazardous chemical release notification, chemical inventory reporting, and toxic chemical release reporting.

Title 40 CFR Part 355 covers state and local structure to prepare for chemical emergencies, emergency release notifications by facilities to local emergency planning committees, and state emergency response. Title 40 CFR Part 370 requires facilities to submit Material Safety Data Sheets (MSDSs). Title 40 CFR Part 372 requires notifying EPA of toxic chemical releases.

Pollution Prevention Act

The Pollution Prevention Act (PPA) has been around since the first of this decade and gives data reporting requirements for toxic release inventory chemicals. It goes beyond hazardous wastes and encourages the maximum possible elimination of all waste types. Volume reduction and reuse are emphasized as the preferred means of pollution prevention.

Resource Conservation and Recovery Act

Subtitle D of the Resource Conservation and Recovery Act (RCRA) of 1976 encourages environmentally sound solid waste management. Of interest to EMS auditors are Title 40 CFR Part 246, which, has a key requirement that office facilities with more than 100 workers separate high grade paper for recycling, and Title 40 CFR, Parts 254–257, which deal with citizens' suits for RCRA noncompliance.

RCRA also allows use of BMPs and provides this general approach for conducting an environmental assessment or audit: (1) study of the management structure that provides oversight and that conducts waste management operations; (2) review of available documentation (agreements, court orders or decrees, program plans, site action plans, regulatory documents of the EPA and state related to specific site activities, and results of previous site audits/assessments/inspections, as well as actual waste analysis records); (3) examination of all major waste management activities under way, with field visits; and (4) interviews of site and regulatory personnel.

Title 40 CFR Part 260 gives the general requirements for a hazardous waste management system. Title 40 CFR Part 261 lists the means of identifying listed and characteristic hazardous wastes. Title 40 CFR Part 262 specifies requirements for hazardous waste generators. Title 40 CFR Part 263 specifies requirements for transporters of hazardous waste. Title 40 CFR Parts 264 and 265 specify requirements for hazardous waste treatment, storage and disposal facilities. Title 40 CFR Part 266 gives the standards for materials being recycled or reused. And Title 40 CFR Part 268 specifies requirements associated with management of land disposal restricted hazardous waste.

RCRA regulates the generation, treatment, and disposal of hazardous waste from cradle to grave and seeks to encourage alternatives to land disposal through recovery of useful material in order to reduce waste volume. RCRA is designed to protect air, surface, and groundwater from contamination due to improper handling of hazardous waste. It was amended in 1984 when the Hazardous and Solid Waste Amendments were passed by congress.

A very active RCRA program means established reporting, record keeping, performance, and operating standards and requirements for all applicable facilities.

Title 40 CFR Part 262 covers hazardous waste manifests to track waste from cradle to grave; packaging, labeling, marking, placarding and on-site accumulation time, record keeping and reporting, and exporting/importing hazardous wastes.

Title 40 CFR Part 265 provides interim status standards for hazardous waste treatment, storage and disposal facilities, and preparing for and preventing hazards; contingency planning and emergency procedures; a manifest system; record keeping and reporting; monitoring; use of containers; and general requirements for waste analysis, security, facility inspection, and personnel training.

Title 40 CFR Parts 271 and 272 outline the procedures by which a state may obtain authorization to implement the RCRA program in lieu of the federal government.

Other related standards, Title 49 CFR Parts 171–180 (Department of Transportation regulations) set forth proper shipping standards for transportation of hazardous waste and waste handling, including the packaging specifications.

Toxic Substances Control Act

The Toxic Substances Control Act of 1976 gives the EPA authority to regulate the manufacture, distribution in commerce, use, disposal, and processing of materials it has identified in its inventory. It also outlines record keeping requirements for manufacturers/importers, processors, distributors, and users of listed chemical substances, including prohibited acts, and describes penalties for violators. The EPA conducts risk assessments using information obtained through the Preliminary Assessment Information Rule (PAIR).

Any person intending to manufacture or import a chemical substance not already listed by the EPA and not prohibited or excluded from TSCA must file a Premanufacture Notification (PMN) to the EPA. If the EPA determines the material's proposed use is a "significant new use" of an existing chemical, then it can issue a Significant New Use Rule which gives EPA 90 calendar days to comment before issuing a Significant New Use Notice.

Other Federal Environmental or Environmentally Related Legislation

Other Acts of lesser interest to an EMS auditor are the National Environmental Policy Act (NEPA) of 1969, Hazardous Material Transportation Act (HMTA) of 1975, and the Federal Insecticide, Fungicide, and Rodenticide Act (FIFRA) of 1979.

Occupational Health and Safety Act

The Occupational Safety and Health Administration (OSHA) is the regulator of the Occupational Safety and Health Act of 1970. The act applies to private sector employers, with the exception of the transportation industry, certain nuclear energy industries, and the mining industry.

Main provisions of the Act include standards to limit exposure to various chemical substances that could induce acute or chronic health effects, regulate substances that may cause cancer, inform employees of dangers via MSDSs (nicknamed HAZCOM), and require employers to maintain medical and other records to track development and incidence of occupationally-induced disease.

As for ISO consideration of occupational health and safety (OH&S), the ISO ad hoc committee has suggested convening a panel to address the overall perspective of OH&S that would describe the OH&S environment and the borderlines between regulatory and voluntary initiatives. This panel is to provide guidance on what is needed in the marketplace to satisfy regulations and requests. They have proposed a workshop limited to management standards to define key aspects of OH&S with an aim to establish:

- the stakeholders
- stakeholder needs
- the value of OH&S management standards for society and their ethical value
- the value of OH&S standards in terms of business performance
- the value of OH&S standards for large, medium, and small businesses
- the differences in perception between countries with high and low levels of OH&S regulation

A cost/benefit analysis on the issue of conformity assessment and auditing of OH&S systems is also planned.

The American Industrial Hygiene Association (AIHA) is leading the way for either a U.S. national or an international standard for OH&S management systems. John Magher, manager of technical support for AIHA, claims that OH&S standards would be a good way for companies to put down on paper their existing good practices. He attended a meeting in Germany in September of 1995 sponsored by the EC to discuss the possibility of including OH&S elements in quality management system requirements. He says that the integration of OH&S standards into either existing QMS or EMS standards is crucial to reach a consensus and that major resistance to the initiative is not whether the OH&S principles are valuable and important to industry, but rather that certification costs could add another burden.

As one TAG member puts it, "With the labor organizations, insurance companies, and academics taking an active interest in OH&S along with different types of industry, reaching a national consensus will not be easy."

Technology Improvement Act

The Technology Improvement Act of 1996, which President Clinton signed into law in March of 1996, espouses some very current ideas in Washington, DC, policy-making circles. Chief among these is increasing use of private-sector voluntary consensus standards, such as ASME codes, ANSI standards, and ASQC guidance documents.

The new law amends the 1980 Stevenson-Wydler Technology Innovation Act and the 1986 Federal Technology Transfer Act, primarily with regard to intellectual property issues.

Federal agencies now must use consensus standards where possible or report their reasons why they do not to the Office of Management and Budget (OMB). The law also requires federal agencies to consult with and participate with voluntary, private-sector consensus standards organizations in developing technical standards. The EPA has been very involved in the drafting of the ISO 14000 standards.

The law mandates that the National Institute of Standards and Technology (NIST) coordinate technical standards and conformity-assessment activities among federal, state, and local governments with private-sector standards activities, with the goal of eliminating unnecessary duplication and complexity in the development and promulgation of conformity assessment requirements and measures.

Incentives for Self-Policing, Discovery, Disclosure, Correction, and Prevention of Violations

The EPA has the new Incentives for Self-Policing, Discovery, Disclosure, Correction, and Prevention of Violations. Awareness of the key points of these incentives is of value for an EMS auditor.

The Incentives are a refinement of the March 1995 Voluntary Environmental Self-Policing and Self-Disclosure Interim Policy Statement of the EPA. The EPA will protect public health and the environment by reducing civil penalties and not recommending criminal prosecution for regulated entities that voluntarily discover, disclose, and correct violations.

"Due Diligence," as defined in the Incentives, reads very much like the ISO 14001 standard. It encompasses the regulated entity's systematic efforts, appropriate to the size and nature of its business, to prevent, detect, and correct violations through all of the following:

(a) compliance policies, standards, and procedures that identify how employees and agents are to meet the requirements of laws, regulations, permits, and other sources of authority for environmental requirements;

(b) assignment of overall responsibility for overseeing compliance with policies, standards, and procedures and assignment of specific responsibility for assuring compliance at each facility or operation;

(c) mechanisms for systematically assuring that compliance policies, standards, and procedures are being carried out, including monitoring and auditing systems reasonably designed to detect and correct violations, periodic evaluation of the overall performance of the compliance management system, and a means for employees or agents to report violations of environmental requirements without fear of retaliation;

(d) efforts to communicate effectively the regulated entity's standards and procedures to all employees and other agents;

(e) appropriate incentives to managers and employees to perform in accordance with the compliance policies, standards, and procedures, including consistent enforcement through appropriate disciplinary mechanisms; and

(f) procedures for the prompt and appropriate correction of any violations and any necessary modifications to the regulated entity's program to prevent future violations.

The EPA will not seek gravity-based penalties or generally not recommend criminal prosecution if the violation results from the unauthorized criminal conduct of an employee and is found through voluntary audits or efforts that reflect due diligence and if the conditions of the EPA policy are met. The EPA will reduce gravity-based penalties by 75% if violations are discovered by means other than audits or due diligence efforts but are promptly disclosed and expeditiously corrected, provided the EPA policy conditions are met.

The policy requires companies to prevent recurrence of the violation and remedy any environmental harm. Repeated violations or violations that may have imminent and substantial endangerment or result in actual serious harm are excluded. Companies remain criminally liable for violations from conscious disregard of legal duties, and individuals remain liable for criminal wrongdoing.

The policy ensures public access to information. EPA may require a description of the company's due diligence efforts be made publicly available as a condition of penalty. Also, EPA requires the company to make written agreements, administrative consent orders, or judicial consent decrees publicly available.

The EPA remains firmly opposed to establishing a statutory evidentiary privilege for environmental audits or blanket immunities for irresponsible conduct and the policy is a positive alternative to privileges that could be used to shield evidence of violations or criminal misconduct, deny the public its right to know, drive up litigation costs, and create an atmosphere of distrust between regulators, industry, and local communities.

Conformity Assessment

A conformity assessment includes all activities that are intended to assure the conformity of products or systems to a set of standards. This can include testing, inspection, certification, quality system assessment, and other activities

A few key points from this definition seem to fit nicely in this section, so that you, as an EMS auditor, will understand that the final definition of conformity assessment has yet to be concluded. In fact, the definition will, the author believes, continue to evolve for quite some time.

In an article by Joel Urman, Program Director of Standards with IBM, he considers that conformity assessment equals assurance. In his opinion, the ultimate value of any standard depends a great deal on how well it is implemented. Conformity assessment is concerned with those activities and procedures that determine if standards are implemented properly. Without conformity assessment, Urman believes, many of the benefits of standards would be lost.

Urman sees conformity assessment as simply confirmation that something does what it is supposed to do. For most things, we do the final assessment ourselves. But, more specifically, conformity assessment encompasses the procedures and activities by which products, processes, services, and management systems are assessed, and the degree to which conformance to a particular standard or specification is met. It focuses on how standards are used and implemented, and how well the supplier's product, processes, etc. meet the specifications defined in the standard.

For EMS standards, conformity assessment is all of the activities needed to determine if and to what extent an organization complies with ISO 14001.

Suppliers, users, purchasers, and government regulators all need assurance that standards are met, according to Urman. Suppliers must satisfy themselves that the standards they use, and those their customers require, are effectively and properly implemented in products, processes, services, and their management systems. Users and purchasers need assurance that they are getting the benefits of the standards in the product, process, or service they buy. Government regulators, mandated to enforce laws and regulations, must have assurance that specified standards are implemented and specifications are met.

Simplified, a product, process, or service must first be tested to determine if it conforms to specification. Tests usually consist of determining specific characteristics of a product, process, or service. Test results are then evaluated to determine if the specifications are met properly.

Management systems have no product to test. Instead, a specially-trained auditor evaluates whether an organization has implemented properly the specifications in its management system.

In a separate article, Urman and Jean McCreary propose that conformity assessment generally takes two forms: third-party assessment for certification or self-declaration of conformity.

Conformity assessment, according to Urman and McCreary, is a time-consuming and costly activity. These costs escalate when the same product, process, service, or management system must be recertified for each country or region.

The ISO organization Committee on Conformity Assessment (CASCO) has as its function the study of conformity assessment of products, processes, services, and quality systems, and the development of international guides relating to testing, inspecting, and certifying products, processes, and services, and relating to assessing quality systems.

There are CASCO guidance documents that provide a common approach for all conformity assessment bodies to follow throughout the world. As of early 1996, the guides did not specifically mention EMS. TC 207 and CASCO are investigating whether new guides need to be written or if current guides should be revised to include EMS elements (ISO/IEC Guide 61, "General Requirements for Assessment and Accreditation of Certification/Registration Bodies," was to be adopted as CASCO 226, Rev. 2; and ISO/IEC Guide 62,

"General Requirements for Bodies Operating Assessment and Certification/ Registration of Quality Systems").

Urman and McCreary conclude that the conformity assessment process provides a mechanism for leveling the field for ISO 14001 implementation worldwide. Lessons learned in the ISO 9000 experience have demonstrated the importance of having mutual recognition of certifications globally. It will be necessary to ensure that accreditation bodies apply similar criteria to the entities being accredited in order to achieve this.

At the domestic level here in the United States, the conformity assessment process should ensure that the customer can have confidence in the organizations implementing ISO 14001. Conformance assessment is simple in concept but may be complex in execution. It encompasses many disciplines and uses the skills and experience of many specialists. When done effectively, it is a valuable and productive discipline that can significantly benefit organizations, users, and government regulators.

An article by Roger Brockway, the Environmental Manager of the UKAS, indicates that companies that develop 14001 EMSs are encouraged to demonstrate their conformity to the standard, either by self-determination or by certification/registration. Independent third-party certification provides the most confidence that the EMS does meet the requirements of the specification.

The last article in the chapter on conformity assessment is from CEEM. It relates that the CASCO guides require that a certification body must be an identified legal entity, either private or public, financially stable, and must maintain the necessary resources to operate a certification system; must satisfactorily document its legal standing and means of financial support; must clearly describe whether it links to a larger entity; must be impartial in judgments; must set up an EMS governing or advisory board; and must stipulate that permanent personnel under the senior executive responsible to the governing board and the executive are free from control by those who have a direct commercial interest in the products/services concerned.

The author believes that there are numerous purposes for standards:

- for commonality
- for consistency
- for efficiency
- for effectiveness
- for reliability
- for safety

Conformance to product standards is best assessed by inspection, test, examination, or use. Conformance to process standards, by audit or assessment.

Conformance to management system standards, by audit, assessment, or review. Another unwritten assessment of conformance is by failure of the product, process, or management system.

Conformance assessment looks for demonstration of conformance and involves verification of objective evidence and determination of degree of conformance. Demonstration of conformance assessment may take the form of certification, accreditation, registration, self-declaration, customer acceptance, public acceptance, or regulatory acceptance.

Management Reviews

Management reviews provide a form of self-assessment to support EMS audits. subclause 4.6 of ISO 14001 requires top management to determine intervals for their reviews of the EMS. The purpose of the review is to ensure the continuing suitability, adequacy, and effectiveness of the EMS. An EMS audit, in the same light, is conducted periodically (predetermined intervals) to determine whether the EMS conforms to requirements and has been properly implemented and maintained. The information from the audit, along with any other information necessary for management to carry out their review, is provided to top management.

Each management review must address the need for any changes to the environmental policy, the environmental objectives, or other elements of the EMS. Changing circumstances and the organization's commitment to continual improvement must be considered by top management in their review.

Let us go back to Principle 5 of the ISO 14004 principles and guidance standard:

"An organization should review and continually improve its environmental management system, with the objective of improving its overall environmental performance."

The standard recommends the management review address all environmental dimensions of the organization's activities, products, and services. Any impact on financial performance or the organization's competitive position should be addressed as well.

The concept of continual improvement is embodied in the EMS, as mentioned throughout this book. Continual improvement is achieved by continually evaluating environmental performance against policy, objectives, and targets to identify opportunities for further improvement.

Continual improvement means finding, correcting, and preventing the root causes of nonconformances and environmental impacts or insults. For corrective or preventive actions to be effective in response to significant conditions adverse to environmental management, the root cause must be investigated, analyzed, and specifically identified.

It is the responsibility of top management to identify the root cause of significant nonconformance to the organization's EMS and see that the causes are eliminated. Although there is no guidance on root cause investigation or analysis in the ISO 14000 standards, there are several methodologies currently in use in the United States and elsewhere, such as the management oversight risk tree, fault tree analysis, effects and causal factors analysis, and others.

Root causes are almost unequivocally found to be management deficiencies. The direct cause of any nonconformance may be different, such as why insufficient personnel resources were allocated regarding achieving set targets, or the verbal directions given to the work force to deviate from a specified procedure. But failures, weaknesses, or assumed risks in the management systems allowed the conditions to be present that resulted in the mishaps. The key to eliminating adverse results is eliminating the root cause before it occurs.

Let us say you, as an EMS auditor, find a worker not following the instructions of the controlled procedure. If the worker has been found competent and adequately trained to perform the procedure, something in the management system governing the work activity has allowed the failure to follow procedure. Perhaps direction from supervision, lack of supervision, errors in the procedure, use of an outdated procedure, or any host of other reasons may be found as the cause.

Management review may also be one form of conformity assessment. It should be a careful and thoughtful look at EMS performance, structure, and implementation. It should be a provoking and enlightening inspection by top management of areas for improvement.

Initial Reviews (see Chapter 3)

Initial environmental reviews may cover a long list of considerations:

- identifying legislative and regulatory requirements directly applicable to an organization's environmental aspects
- identifying environmental aspects and those of significance in potential impact or liability
- evaluating environmental performance
- evaluating existing management practices and procedures
- obtaining feedback from previous legislative noncompliance investigations
- providing opportunities for competitive advantage improvements
- obtaining views of interested parties, internal and external
- evaluating impediments to environmental performance

Note that these considerations are similar to ones the EMS auditor takes into account in audit planning, audit agenda development, assignment of audit team personnel, and even audit checklist preparation. An initial environmental review may be one of the most useful activities an organization should undertake before documenting its environmental policy or its environmental objectives and targets. The review is to assess where the organization is in relation to predetermined criteria.

Let us revisit three important definitions that apply to initial environmental reviews:

- **Environmental aspect**: Element of an organization's activities, products, or services which can interact with the environment.

- **Environmental impact**: Any change to the environment, whether adverse or beneficial, wholly or partially resulting from an organization's activities, products, or services.

- **Environmental performance**: Measurable results of the EMS, related to an organization's control of its environmental aspects, based on its environmental policy, objectives, and targets.

Any interaction with the environment is an environmental aspect—breathing in and breathing out, driving an automobile, burning leaves, going fishing. When the interaction has or can cause a significant change to the environment, it is labeled a significant environmental aspect, whether the impact is beneficial or adverse.

The initial environmental review is used to identify the environmental aspects; significant environmental impacts, whether actual or potential; and the significant environmental aspects, those that change the environment significantly. Chernobyl, Bhopal, Jonestown: the changes were significant and adverse. Elimination of chlorofluorocarbons (CFCs) for air conditioning was a significant beneficial environmental change or impact. Seeing environmental impacts only as adverse means you have not taken off your compliance hat!

For the environmental policy, an initial environmental performance review may be useful. Key areas to review include:

- legislative and regulatory requirements (are they identified?)
- environmental aspects with significant environmental impacts and liabilities (are they identified?)
- significant environmental issues (are they identified?)
- performance against internal criteria
- performance against external criteria

- existing management practices and processes (gap analysis?)
- procurement and contracting policies and processes
- feedback on previous nonconformances and noncompliances
- views of interested parties (customers included)
- impediments to environmental performance/improvements

Some environmental aspects to consider during an initial review may be:

- controlled and uncontrolled emissions
- controlled and uncontrolled discharges
- solid and other wastes (waste streams)
- contamination of land
- use of land, water, fuel, energy, and natural resources
- noise, odor, dust, vibration, and visual impact

The associated environmental impacts and their significance are then determined. Those with significant impacts are significant aspects. From the review, the organization will have more knowledge of its environmental performance on which to base its environmental policy, environmental objectives, and environmental targets.

With regard to environmental performance, let us add the word "evaluation" to the term. As specified in ISO 14001, 4.5.1-Monitoring and measurement, the organization shall establish and maintain a documented procedure for periodically evaluating compliance with relevant environmental legislation and regulations. It shall monitor and measure on a regular basis the key characteristics of its operations and activities that can have a significant impact on the environment. It shall record information to track performance, relevant operational controls, and conformance with objectives and targets.

The ISO 14030 standards are for environmental performance and define environmental performance evaluation as an ongoing review process conducted by line people as opposed to an independent audit.

Remember, in an environmental review or environmental performance evaluation, it is important to understand the inputs and outputs to the environmental aspects and environmental impacts, the competitive advantage to be gained by environmental performance evaluation, the commitment to continual improvement, and the need for resources—time, money, people, training, tools, communications, and authority.

Life-Cycle Assessment (LCA)

For LCA, there are the ISO 14040 standards. They define LCA as a systematic set of procedures for compiling and examining the inputs and outputs of materials and energy and the associated environmental impacts directly attributable to the functioning of a product or service system throughout its life cycle, from "cradle to grave."

An LCA is defined as the compilation and evaluation of the inputs and outputs of materials and energy and the potential environmental impacts of a product system throughout its life cycle. The assessment is to be done in accordance with a systematic set of procedures. During a LCA, there is the compilation and quantification of inputs and outputs of a given product system, known as "life-cycle inventory analysis." This involves "life-cycle valuation." Within the assessment, there is an element called "life-cycle classification" in which the inventory parameters are grouped together and sorted into a number of impact categories. There is also the element of "life-cycle characterization" in which the potential impacts associated with the inventory data in each of the selected categories are analyzed. From these elements of the assessment comes "life-cycle impact assessment," the understanding and evaluating of the magnitude and significance of the potential environmental impacts of a product system. The synthesis drawn from either the inventory analysis or the impact assessment, or from both, is done in line with the defined goal and scope of the assessment and is called the "life-cycle interpretation."

A product's life cycle, as with a process or system, has a beginning, middle, and end. At any point along this line, there are potential environmental impacts. It starts usually with the extraction of raw materials for processing, goes through delivery and processing of the materials, the use of energy during production, the final product and its transportation to consumer, and its end disposal.

The RCRA legislation has a similar cradle-to-grave philosophy, but is limited to hazardous materials. The philosophy and scope of ISO 14040 and LCA are far broader in scope and application.

Members of TC 207 Subcommittee 5 on LCA have long been struggling to harness the intricate concepts that are now more than 20 years in the making and still fraught with tangles. The Subcommittee hopes the four standards (14040, Principles & Practices; 14041, Inventory Analysis; 14042, Impact Assessment; and 14043, Interpretation) will be practical, useful, and economical. At its heart is a struggle to analyze and predict the behavior of complex systems in a world where scientific data exceeds scientific understanding at every turn. Practical application of an LCA is spotty and generally has been product-specific, such as soda beverage containers. The subcommittee's mission remains to address LCA on a systems level.

The LCA standards are a reaction by ISO to the worldwide recognition that consumption of manufactured products has placed a great demand on the environment and the earth's natural resources. These demands are affected by every stage of a product's life cycle and the web of consequences that radiate from each stage.

It is understood by the author that LCA has become one of the most actively considered techniques for studying strategies to meet environmental challenges. The basic objective is to guide decision makers, such as industrialists and policy makers, as well as consumers, in selecting actions to minimize environmental impacts. To do this, LCA must work with traditional motives for selecting one action over another, motives impelled by economic and social goals. LCA also requires that its users consider the products and processes as well as the tools and systems that gave life to the product. This is where a good EMS can play a significant role in identifying what is best for the local environment and the global environment.

Remember the definition of environment: the surroundings in which an organization operates. It includes air, water, land, natural resources, flora, fauna, humans, and their interrelation. It extends from within the organization to the global system.

History

The first LCAs were conducted in the late 1960s and early 1970s when energy efficiency, recycling, and solid waste were issues of public concern. At least two such assessments were performed in the late 1970s and compared alternative products. Other studies focused on the energy and environmental impacts of packaging alternatives, such as disposable versus reusable diapers.

The Coca Cola Company, in 1969, funded a study to compare different beverage containers and to determine which container released the fewest contaminants to the environment and consumed the fewest natural resources. This later became known as the Resource and Environmental Profile Analysis (REPA). A similar tool named Ecobalance was developed in Europe in the early 1970s.

The EPA, in 1974, commissioned a follow-up study on beverage containers. This is often referred to as the study that marked the rebirth of LCA. But with the fading of energy issues in the mid-1970s, so did interest in conducting LCA, energy-oriented studies.

In the mid-1980s, a directive on food containers issued by the European Economic Community charged companies with monitoring the energy and raw materials consumption and solid waste generated by their products. LCA emerged as the tool for analyzing the environmental impacts of products once again. Now, the Society of Environmental Toxicology and Chemistry (SETAC) is playing a leading role in bringing practitioners and users together in developing LCAs.

In 1990, SETAC organized a workshop, "A Technical Framework for LCA," with the goal of identifying the state of the art in LCA and areas where research was needed. The workshop resulted in a book presenting the framework. SETAC divided LCA into three main areas: inventory analysis, impact analysis, and improvement analysis.

LCAs, to date, have been performed on products of low complexity. But consider a more complex product like the personal computer. Housed inside are some 750 or more materials and chemicals that got there through millions of process steps. An inventory would produce billions of data points. How can decisions be made when new product introduction cycle times are between 6 and 18 months versus a design cycle of many years for a beverage container with three elements? LCAs tend to focus on resource consumption, emission, health effects, and ecological impacts. They do not evaluate the cost of possible alternative strategies under consideration. No policy can be evaluated without understanding the economic implications of the alternatives.

LCA methodologies are currently based on encyclopedic volumes of data collection, in minute detail, which include every material, resource, process, and emission in the product or process under analysis. An immense amount of data is produced, and this makes it impossible to evaluate or reach a product decision. Even choosing which data to analyze can be overwhelming.

Given that the global regulatory process changes much more rapidly than our new product introduction cycle, the question is, "Does this process have a greater significant impact than our present quality/process control methodologies?"

Process

The LCA is a tool to collect and interpret information for a variety of purposes. The information may be of assistance to anyone involved in decision making, selecting relevant environmental performance indicators for environmental performance evaluation (discussed later in this chapter), or eco-labeling and marketing.

LCA is a systematic tool for assessing environmental impacts associated with a product or service system. It has as its goals building an inventory of inputs and outputs, making a qualitative and quantitative evaluation of the inputs and outputs, and identifying the most significant environmental aspects of the system.

LCA encourages organizations to integrate environmental issues into their overall decision-making process. But despite the fact that LCA involves mainly quantitative processes, there exist certain limitations. Information regarding environmental, manufacturing, distribution, and waste management conditions may change and vary from region to region. It is neither possible nor practical to always collect a complete data set directly relevant to the subject of a LCA. The scope, boundaries, and detail of a LCA will also depend on the use of the study.

Any LCA, regardless of scope and level of detail, will involve assumption, value judgements, and trade-offs. It must be clearly communicated with unsubstantiated claims avoided. Great caution should be used in making environmental claims based on LCA results. The ISO 14040 standard establishes normative requirements of comparative assertions of products that are disclosed to the public.

Decision making should not rely solely on LCA as the only tool or source of information. Decisions should typically involve a number of other important factors such as risk, benefit, cost, and other environmental issues, including the views of interested parties.

LCA is still in its infancy. If it is to gain widespread acceptance and enhance overall environmental performance, it must maintain its technical credibility while being flexible, practical, and cost-effective when applied so that it is consistent with the differing goals and needs of relevant stakeholders. This is especially true for small- to medium-sized enterprises.

Environmental Performance Evaluations (EPE)

An EPE measures, analyzes, assesses, and describes an organization's environmental performance against management defined criteria for various reasons. The principles of an EPE depend on the organization's purpose, nature, and extent of its relationship with the environment, its ability to control or influence its relationship with the environment, and its ability to measure effects. Organizations should focus on their significant environmental aspects. Measurement for measurement's sake should be avoided. Significance can be determined by analysis, political interpretation, commercial judgement, technological means, or social measures.

The definition of environmental performance is the sum of measurable outputs of the EMS relating to an organization's control of the environmental impact of an organization's activities, products, or services. It is based on environmental policy, objectives, and targets. Environmental performance indicators provide a means of evaluating and describing the environmental performance achieved by an organization.

Process

The EPE process should be useful, as a management tool, and user-friendly—easy to understand and fairly transparent. It should be compatible with the EMS and assist in attaining the environmental policy, objectives, and targets. It should be selective, concentrating on those environmental performance indicators that are truly related to significant environmental aspects. And, of course, it must be verifiable.

The EPE has three primary components: (1) management systems, (2) operational systems, and (3) the environment. Its benefits to an organization may be any or all of the following:

- provide better understanding of environmental impacts
- contribute to the ongoing identification and prioritization of environmental policy, objectives, and targets
- assist in aligning activities among organizational units
- provide evaluation of environmental risks and liabilities
- measure financial gains or costs associated with environmental management programs
- serve as a basis for continual improvement
- serve as a rationale for resource allocations
- serve as a basis for performance improvements and incentives
- provide measurement, assessment, and analysis of performance against policy, objectives, targets, and improvement programs
- provide self-assessment and comparison of performance to legal and regulatory requirements
- aid in identifying root causes of performance deficiencies
- establish leading preventive indicators for possible future performance deficiencies
- serve to demonstrate achievement
- serve as a basis for communication to internal and external stakeholders

Environmental Site Assessments (ESA)

The goal of an ESA is to identify recognized environmental conditions. These conditions indicate the presence or potential presence of hazardous substances or petroleum products on a property under conditions that indicate an existing, past, or the threat of a release. The release in this context may be into structures, the ground, groundwater, or surface water on the property.

There are accepted practices for ESAs, in particular, ASTM Practices E 1527 and E 1528. The first is a procedure for an ESA called the "Phase I Environmental Site Assessment Process." Its companion is a "transaction screen process" to qualify a buyer of a property under the innocent landowner defense clause of CERCLA. The transaction screen process may be conducted by the user or for the user by an environmental professional, in whole or part.

The Phase I site-assessment must be performed by an environmental professional with errors and omissions insurance.

If the transaction screen process is determined by the user to conclude that no further inquiry into previous ownership and uses of the property is needed to assess the potential for identifying any recognized environmental condition, no further action is taken. Performance of the transaction screen process constitutes the "appropriate inquiry" required by CERCLA for innocent landowner defense. If the conclusion by the user is that further inquiry is needed, the user must decide whether it may be limited to those specific issues identified of concern or if a full Phase I assessment is necessary.

Risk Assessment

- Personnel qualified and experienced
- Data needs for assessments
- Data collected
- Selection of contaminants
- Human health exposure risk assessments
- Human health toxicity risk assessments
- Human health risk characterization
- Environmental risk assessment techniques
- Environmental risk assessment study results
- Formal documentation

✿Personnel qualified and experienced✿

Personnel must be qualified and experienced to conduct and oversee risk assessments. They should either be on the organization's staff or subcontracted as a service.

✿Data needs for assessments✿

In human health and environmental risk assessments, data needs are evaluated early in the scoping process. Data elements required are then specifically addressed in sampling and analysis plans.

✿Data collected✿

Data for risk assessments are collected using appropriate techniques to provide the required detection limits prior to conducting the risk assessment.

✿Selection of contaminants✿

Contaminants of concern are selected. The selection process is fully justified. The selection of contaminants is documented.

✿Human health exposure risk assessments✿

For human health risk assessment, exposure assessments are carried out on relevant exposure pathways. Exposures to sensitive receptors are determined through quantitative analysis, then documented, reviewed, and approved by appropriate parties.

✿Human health toxicity risk assessments✿

For these human health risk assessments, the latest valid toxicity information available is utilized. The values are based on the hierarchy of acceptable EPA sources or those approved by other authorized parties (SETAC, for one).

✿Human health risk characterization✿

These human health risk assessments use risk characterization to present baseline risk results. The considerations involve a list of significant assumptions and uncertainties associated with the risk results.

✿Environmental risk assessment techniques✿

Techniques involve studying and assessing all potential impacts of site releases on biological receptors. Remember that, under ISO 14001, the environment is from within the organization to the edges of the world.

✿Environmental risk assessment study results✿

Study results address the current observed effects. They are used to summarize the risks and the threats to affected populations.

✿Formal documentation✿

Formal documentation is maintained to support the facility's risk management strategy at its hazardous substance release sites or to support the facility's decision on risk acceptability. To support the facility's overall risk management program, routine site-wide surveillance or monitoring of release sites is conducted.

15   EMS Auditor Qualification, Certification, and Training

EARA and the RAB

The author has chosen to provide information regarding the environmental and/or EMS auditor certification criteria from two of the most widely recognized bodies to date: the Environmental Auditors Registration Association (EARA) and the Registrar Accreditation Board (RAB).

One of the first things that must be considered is the type of certification you wish to apply for and ultimately achieve. The differences at this point in time between the EARA and RAB schemes are night and day. However, both schemes do share an obvious common interest—their dedication to raising standards of professional competence in the environmental auditing profession. Before continuing, it is critically important to note that the references made in this chapter are from the RAB certification criteria for EMS auditors. It should be noted, of course, that any comparisons and/or differences illustrated in this chapter are subject to change, including any modifications, updates, and/or improvements made within the EARA scheme in 1997.

There is a great demand globally for both schemes. The EARA scheme is considerably more in line with the advancement, recognition, and competencies of a broad range of environmental audit disciplines. The following qualification experience criteria are contained within Appendix I of the EARA registration scheme and application form.

A1   Conducting preparatory review of existing environmental performance, policies, or management practices

A2   Formulating or writing of company environmental policy/ statement (either for internal use or public issue)

A3   Formulating of quantified environmental targets and objectives as the basis for subsequent audit

295

A4   Writing of environmental management manuals or formulating environmental management programs/strategies

A5   Coordinating implementation of environmental management systems (other than for BS7750 certification)

A6   Coordinating/managing BS7750 implementation

A7   Coordinating/managing eco-audit implementation

A8   Performing Environmental Management System audit (internal)

A9   Performing audit for certification to BS7750

A10   Verifying company audit findings or environmental audit statements for eco-audit or other purposes

A11   Performing multi-issue site audits against corporate environmental policy or environmental legislation

A12   Performing single-issue energy or waste minimization auditing

A13   Performing "Green" product audits/cradle-to-grave/life-cycle analysis

A14   Performing risk assessments, environmental liability, or due diligence audits for companies involved in preacquisition, divestiture, or merger activities

A15   Running staff training seminars, work groups, or lecturing at conferences on audit-relevant topics

This list is not exhaustive. Specific technical or other audit experience should be described in full if not covered by any of the above categories.

The EARA scheme takes into account and recognizes multi-issue disciplines in the environmental arena. The scheme, at this point, is well-suited not only for EMS auditors, but for a strong contingency of environmental regulatory compliance auditors as well.

From the RAB certification criteria for EMS auditors:

The auditor and the auditor's organization shall have independent management and operating structure from the audited organization. Examples of acceptable relationships are:

- *A head office audit of a plant or division*
- *One division or plant auditing another division or plant*
- *A customer organization auditing a supplier*
- *A third-party registration audit*
- *A consultant contracted to provide an independent conformance audit*
- *An audit of a registrar by an accreditation body*

All audits conducted in accordance with ISO 14011, including full-system audits, partial-system audits, and surveillance audits, are eligible. (See also Section 3.4 Environmental Auditor).

In the author's view, it will certainly be interesting to see how the RAB verifies and validates that the applicants, for their "relevant" audit experience, were conducting EMS audits in accordance with ISO 14011. In addition, what "equivalent" corporate, organizational, and/or nationally recognized EMS standard will be regarded as equal to ISO 14001?

Application Structure and Content

The following outline depicts the type of information required to be submitted for application to the EARA scheme of registration. Many of the same types of information will also be required for application to the RAB.

1. Summary of Technical and Academic Qualification
    A. Year completed
    B. Course length
    C. Course title
    D. Award/grade
    E. Educational establishment
    F. Description of relevant subject areas studied
2. Membership of Professional Bodies
    A. Professional body
    B. Date of joining
    C. Present membership level/qualification
    D. Description of relevant structured training and exams undertaken (includes information regarding how your position within the institution was achieved)
3. Summary of Relevant Training to date
    A. Date and length of course (days)
    B. Assessment or examination completed
    C. Title of course and course organizers or details of on-site training
    D. Subject/topics covered
    E. Means by which this information can be verified
4. Summary of Relevant Experience to date
    A. Date and duration in days
    B. Type of work undertaken (reference Appendix 1 codes)
    C. Industrial sector (reference Appendix 2 codes)

     D. Additional details on role/responsibility (i.e., audit team member, lead auditor, project manager, etc.)

     E. Method by which this information may be verified (name, address and telephone number of employer, client, colleague, or supervisor)

5. Other information to support the application, including declaration, fees, and details for inclusion on the public register

Both EARA and the RAB share the same appropriate background/work experience within their certification criteria, including work experience that contributes to the development of skills and understanding in some or all of the following:

- environmental science and technology
- technical and environmental aspects of facility operations
- relevant requirements of environmental laws, regulations and related documents
- environmental management systems and standards
- audit procedures, processes and techniques

The application requires attention to detail specific to each of the line item requirements listed above for either scheme. The author especially wishes to draw your attention to two critical factors and elements of the evaluation process of the application.

1. **DETAIL•DETAIL•DETAIL!!!**

    Carefully review and identify all information provided in the application. There is no such thing as too much detail, but certainly a concern arises if insufficient detail is included to any criteria requirement.

2. **Means by which this information can be verified.**

    This may be the most important consideration within the application and evaluation process. Submit a statement directly from a supportive manager, project or team leader, and/or supervisor that can attest to the validation of the information provided relevant to the area of inquiry.

Qualification Training

As stated in this chapter, the auditor's best option of choice may clearly lie with an accredited training course provider through both the RAB and EARA. This will allow for the qualification training to be recognized equally by both schemes. An auditor may consider applying to both schemes due to the differences in scheme requirements and the diversity of their environmental auditing experience. However, it is worth noting that, in the author's opinion, multiple certifications defeat the purpose of mutual recognition of the various certification schemes established now and to come.

Most accredited course providers offering lead auditor courses are going to cover a lot of the same material in the internal EMS auditor course. However, internal auditors should consider the lead auditor course as much, if not more than, third-party auditors. Internal auditors will need to ensure the EMS works effectively and meets ISO 14001 requirements…an integral part of preparing the organization for the certification audit process initially. Remember, the best auditee is a qualified auditor.

Organizations should consider sending some internal auditors and environmental engineers to an internal auditor course (especially those offered by accredited training course providers). A few of these—including the management representative and environmental compliance management—should consider the EMS lead auditor course. Additional information of training and qualification of EMS auditors is provided in Chapter 1.

Summary

It certainly goes without saying that certain elements of the certification criteria may, in fact, change or have changed already as you are reading this. First off, let's make available to you, the reader, information on contacting both EARA and the RAB for their current, updated certification/registration criteria:

The Environmental Auditors Registration Association (EARA)
Welton House
Limekiln Way
Lincoln, U.K.
LN2 4US
tel: 01522 540069; fax: 01522 540090
contact: Ruth Bacon/Sharon Milnes

Registrar Accreditation Board (RAB)
P.O. Box 3005
Milwaukee, Wisconsin 53201-3005
tel: 1-800-248-1946; fax: 1-414-765-8661
contact: Barbara Stranek/Janet Jacobson

The most important question to ask yourself with respect to the EMS and/or environmental auditor certification programs is which scheme best suits the needs of the organization, customer base, stakeholders, and relevance to the type of environmental audit experience to date? Likewise, to you, the EMS auditor, which scheme best suits your needs and expectations in professional advancement, recognition, and certification?

So, ask yourself again, "Do I really want to be an EMS auditor?"

ISO 14000
Bibliography and Resource Center

Inside ISO 14000:
The Competitive Advantage
of
Environmental Management
by Donald A. Sayre
St. Lucie Press, 1996

From CEEM Information Services
10521 Braddock Road
Fairfax, VA 22132
Tel: 800-745-5565; Fax: 703-250-4117

ISO 14000 Questions & Answers
All of your questions about ISO 14000 standards are answered clearly and concisely in over 40 pages of reference. Published by CEEM Information Services with ASQC.

The ISO 14000 Handbook
Over 750 pages, including the text of ISO/DIS 14001 and ISO/DIS 14004, provide you with detailed explanations and analysis of all the ISO 14000 series of standards, focusing on practical guidance on ISO 14001 implementation and certification. Edited by U.S. TAG to TC 207 chairman, Joseph Cascio. Published by CEEM Information Services with ASQC.

ISO 14000 Case Studies
Over 200 pages of ISO implementation and auditing case studies from North American companies offer you insight straight from the source.

The ISO 14000 Resource Directory
Locate the EMS resource providers your company needs when you need them! Comprehensive and user-friendly listings provide you company names, mailing addresses, contacts, telephone and fax numbers, E-mail and Website addresses, services offered, client lists, and more.

International Environmental Systems Update
CEEM's newsletter of record on ISO 14000 keeps you abreast of the latest environmental management systems standards developments. An essential information source used by hundreds of companies worldwide.

Index

Results
  appropriate, sufficient 151
  of audits and reviews 140, 145
Retention times 141, 142
Retrievability 142
Reusable audit procedures 179
Reusable data 166, 170, 180
Review 44, 48, 74, 85, 87, 88, 91, 120,
      170, 180, 209, 211, 216, 217, 223,
      230, 264
  proper method of 141
Reviewers, roles of 47
Rio Declaration 46
Risk 3
  assessment 269, 270, 292
  characterization 293
  techniques 293
Roles, responsibilities, and authorities 65,
      92, 96, 97, 101, 110, 113, 170, 217,
      229, 266
Root cause analysis 13, 136, 265
Root causes 283, 284, 291

S

Safe Drinking Water Act (SDWA) 273
Safety precautions 202
Sampling 26, 81, 90, 92, 122, 214, 218
Schedule log 190
Scheduling of audits 145, 181
Second
  audit 56
  shift 196
Second-party
  auditors 179, 185, 230, 241, 257, 267
  audits 190, 267
Selecting audit team members 187
Selection criteria, facility for audit 185
Self-assessment 35, 145, 283, 291
Self-declaration 61
  of conformity 281
Self-determination 282
Shareholder groups 3
"Show me"
  approach 25
  attitude 20
  questioning technique 228

Shut down 74
Significant
  environmental aspects 76, 78, 82, 85,
      87, 99, 104–106, 117, 120, 121,
      129, 132, 140, 232, 238, 285, 289,
      290
    external communication 102
  impacts 77
    monitoring 128
  New Use Rule 277
Site
  emergency plans 107
  management reviews 102
  -specific protocol 166
  tours 221
Society of Environmental Toxicology and
      Chemistry (SETAC) 288
Software and hardware sampling 130
Span of control 103
Spill Prevention, Control, and Counter-
      measures (SPCC) 273
Stakeholders 13, 291
Standard
  assessment questions 232
  Industrial Classification Code 17
Standards Conformance Registrar
      Advisory Group 16
Start-up 74, 134, 141, 175
Stevenson-Wydler Technology Innovation
      Act 278
Stickers 24
Straightforward questions 227
Strategic Advisory Group on Environment
      (SAGE) 3, 16, 29
Strategic plan 88, 171
Strengths and weaknesses 217
Subcommittee 17
Subcontractor 14
SubTAG 17
Subworking group 17
"Suitability" 151
Summary report 242, 250
Superfund Amendments and Reauthoriza-
      tion Act (SARA) 274
    Title III 275
Suppliers 14, 52, 75, 209, 211